Enjoy Writing your Science Thesis or Dissertation! 2nd Edition

A step-by-step guide to planning and writing a thesis or dissertation for undergraduate and graduate science students

Enjoy Writing your Science Thesis or Dissertation! 2nd Edition

A step-by-step guide to planning and writing
a thesis or dissertation for undergraduate and
graduate science students

Elizabeth Fisher
University College London, UK

Richard Thompson
Imperial College London, UK

Imperial College Press

ICP

Published by

Imperial College Press
57 Shelton Street
Covent Garden
London WC2H 9HE

Distributed by

World Scientific Publishing Co. Pte. Ltd.

5 Toh Tuck Link, Singapore 596224

USA office: 27 Warren Street, Suite 401-402, Hackensack, NJ 07601

UK office: 57 Shelton Street, Covent Garden, London WC2H 9HE

Library of Congress Cataloging-in-Publication Data
Fisher, Elizabeth.
    Enjoy writing your science thesis or dissertation! : a step-by-step guide to planning and writing a thesis or dissertation for undergraduate and graduate science students / Elizabeth Fisher, University College London, UK, Richard Thompson, Imperial College London, UK. -- 2nd edition.
        pages cm.
    Previous edition by Daniel Holtom.
    Includes bibliographical references and index.
    ISBN 978-1-78326-420-9 (hardcover : alk. paper) -- ISBN 978-1-78326-421-6 (pbk. : alk. paper)
    1. Technical writing. 2. Dissertations, Academic. I. Thompson, Richard (Richard Charles), 1955–
II. Holtom, Daniel. Enjoy writing your science thesis or dissertation! III. Title.
    T11.H582 2014
    808.06'65--dc23
                                                                                        2014019115

British Library Cataloguing-in-Publication Data
A catalogue record for this book is available from the British Library.

Typeset by Stallion Press
Email: enquiries@stallionpress.com

Printed in Singapore

Acknowledgements

This book is based on a previous version by Daniel Holtom and Elizabeth Fisher and we wish to thank Dan for generously giving us a free hand in updating this second edition. We thank Kellye Curtis from Imperial College Press for encouraging Elizabeth to agree to revise the book and persuading Richard to join her. In addition we thank Tom Stottor for his patience, advice and professionalism during the editing process. We also wish to thank Zara Holtom for creating a much better set of illustrations than were in the first edition ...

We thank Frances Wiseman and Rosie Bunton-Stasyshyn for scientific input to the book and Rohit Jaggi for advice about the use of English.

Elizabeth would also like to thank Edith Sim, Clare Stanford, Alan Kingsman while at Oxford University and David Page of MIT for teaching her about scientific writing. She would like to thank her colleagues at Imperial College for input while writing the first edition of this book and at UCL for the second edition, and David Housman at MIT and Peter Jackson at Stanford University for hosting her sabbatical during which part of this book was produced. Richard would like to thank his wife Margaret for all her support while writing this book and his colleagues at Imperial College for their encouragement and advice.

We would both like to thank all the graduate and undergraduate students who have been through our groups for all their patience and everything they have taught us.

Finally, we are also indebted to our own supervisors (Steve Brown, Mary Lyon and Derek Stacey) who set us excellent examples of effective research supervision.

Contents

Chapter 1 provides basic information and summarises how to organise, plan and write a thesis or dissertation. The remaining chapters go into much greater detail on each topic; because each chapter can be read independently of the main text you will find repetition between them. Use what is helpful to you.

We recommend that you read Chapter 1 as a whole, and then use the rest of the book as a reference section.

Crest of the University of West Cheam (which may, or may not, exist), where the undergraduate and postgraduate students depicted in the illustrations throughout this book are based. The university, formerly the West Cheam Institute of Technologickal Arts, boasts some unconventional academic staff including Dr Karloff, Dean of Alternative Physics, and it carries out research on exotic subjects such as the mating behaviour of landfish. As described in this book, students at the university have encountered many problems, such as failed experiments, crashed computers, lost references, broken printers and errant supervisors. However, despite this, all the people mentioned managed to finish writing their dissertations and theses on time and successfully completed their degrees.

Chapter 1

Overview

You probably have more than enough reading to do and possibly very little time, so this chapter is a short guide to how to write a science thesis or dissertation, with information summarised under useful headings.

Chapter 1 is meant to be read as a whole and will give you a basic understanding of the standard conventions of dissertation and thesis writing, along with suggestions as to the best way of approaching the task. It will help you streamline the process of producing an effective and readable dissertation, thesis, or yearly report for a BSc, BA, BEng, MSci, MPhys, MChem, MSc, MPhil, MEng, PhD, DPhil – or any other degree.

The rest of the book provides more detailed advice about the different aspects of thesis and dissertation writing. Read the parts that are useful to you, take what is helpful and leave what is not.

1.1 Preliminaries

What are the rules?

There are very few rules for writing a dissertation or thesis. The rules that do exist mainly concern the formatting of your work (number of copies, layout, Title Page, Abstract or Summary, binding and so on) and these differ between departments and universities. Find out your local rules immediately – look online and ask your supervisor. Also check whether there is a deadline for the submission of your thesis.

While there are few rules, there are many conventions – modes of writing that we all conform to, usually because they are efficient ways of conveying information. Each discipline has its own conventions for the structure of theses and dissertations. The best way of finding out about the conventions in your area of research, and your department, is to ask your supervisor to recommend a good recent dissertation or thesis in your own field as a guide to form, length, content and style.

Interim reports

You may well have to submit interim reports for your project, for example, quarterly or first-year reports. Approach them in the same way as your dissertation or thesis, although they will be mainly concerned with Materials and Methods, and Results sections. Try to produce a professional and well-written document. These reports might well be usable as parts of your final dissertation or thesis, which will save you a lot of work at the end of your project.

Where to write

Writing your dissertation or thesis will be hard enough at the best of times. Find somewhere to write where you can concentrate on your work and will not be interrupted too often. You will be spending a lot of time in this place so make it comfortable. Your concentration will not be helped if you develop chronic backache, so check your chair is the right size and the table or desk the right height for you. Make sure you have easy access to cups of tea or coffee, biscuits, etc. – the essentials that will keep you going (see Figure 1.1).

1.2 Creating a Plan for Your Thesis or Dissertation

The importance of planning

A dissertation or thesis is not simply a list of experiments or calculations accompanied by a vague outline of what they all mean. The text needs a clear structure that starts by introducing the reader to the

Figure 1.1 The essentials of thesis writing.

topic, then states the aim of the research, shows the results and finally discusses their significance. You need a plan. Without one, it's easy to overlook important points or jump about randomly from idea to idea.

This guide is written on the assumption that you will **first develop a complete plan of your thesis or dissertation**. Only when this is in place and you have defined you aim and can see the overall structure of your text, should you start writing. We can't over emphasise just how important it is to have a plan before you start writing. So we'll put it in capitals and bold: **CREATE A COMPLETE PLAN OF YOUR THESIS OR DISSERTATION BEFORE YOU DO ANY WRITING.**

Structure

In your thesis you are taking your reader on a journey (see Figure 1.2). Starting from an overview of your field, you then need to focus in on the topic that you have worked on, explaining as you go why it is important. After you have described your research methods and your results, you then need to zoom out again, explaining the significance of your work in the broader field and the implications it will have for future studies.

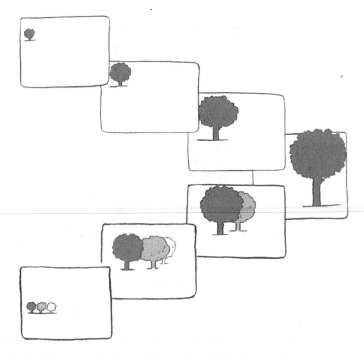

Figure 1.2   Plan your thesis or dissertation. Take the reader on a journey from viewing your field as a whole, to looking at your individual contribution, to placing you work back in the context of your field.

As with any other story, the structure of a scientific thesis or dissertation has three parts: the beginning, the middle and the end. In its final form, your dissertation or thesis will probably be laid out something like this:

The beginning:

Title Page
Abstract
Dedication
Acknowledgements
Table of Contents (including Appendices)
List of Figures
List of Tables

List of Abbreviations (also known as 'Nomenclature' in some disciplines)
Introduction (including a Literature Review)

The middle:

Materials and Methods/Experimental Techniques
Results

The end:

Discussion and Conclusion
References (also known as 'Bibliography' in some disciplines)
Glossary
Appendices
Published Papers

You probably will not have all these sections in your thesis or dissertation, but this scheme provides a basic structure from which to plan your writing.

When planning your thesis or dissertation, take care to refer to each item in order. Do not (as we sometimes see) refer to an item, such as a finding or a method, in one part of your thesis, but describe and explain in detail what this item is *later* in the thesis.

A useful first exercise is to have a careful look through a few recent theses in your field. Browse through them to see how the work has been divided into sections. Look at the layout, formatting, and font. Decide what you like and what you do not like.

Get your information into a workable form

To start making your plan, you need your information (ideas, calculations, data, etc.) in a form that is easy to arrange and rearrange. Decide on the key points for each section of your thesis or dissertation. You could note the points on paper, on your computer or write them on index cards. Once you have your key points, shuffle them about until you are happy with the order. At this stage it's also a good idea to plan roughly the length of each section of the thesis. This helps you to get a good balance between the lengths of the different parts of the document. You can always make changes later. As a very rough: your introductory section; your methods section; your results section; and your discussion/conclusion section should all be roughly of an equal length. This rule is not set in stone but gives a good starting point for getting the right balance between the different sections.

One sensible approach is to make notes for your project in the following order:

- **Literature Review.** This chapter tells the reader the background to your topic and reviews previous work in the area. It will be based on the collection of original sources of information that you gathered during your project. For some disciplines this information is incorporated into the **Introduction**.
- **Materials and Methods** (or some other equivalent title). This chapter of your thesis covers what equipment, techniques and procedures you used and how you used them.
- **Results.** This states what experiments you carried out and what you learnt from these experiments. Some or all of your results may come from computer simulations or theoretical analysis.
- **Discussion and Conclusion.** This should relate the significance of your findings to your field of study, and give your conclusions and suggestions for future research.
- **Introduction.** This introduces the reader to your field of research, your aims, and the experimental system or theoretical framework in which you have been working.
- **Abstract.** This is a condensed version of your whole dissertation or thesis.

Planning your project in this order helps you to see what you did (Materials and Methods), what you have achieved (Results) and what you have learnt (Discussion and Conclusion) – which is often different from what you set out to do and achieve. It then helps you to re-evaluate your original aim (Introduction), and to modify that aim if you have not achieved it, so you can place your results and conclusions in the best context.

The plan given above relates most directly to an experimental project, but it's also applicable to other types of work, including theoretical projects, computer-based projects such as simulations and observational studies such as in astronomy. In all cases it is possible to identify the techniques you used and the results you achieved, whatever form those results take. Throughout this book, when we say 'experiments', we generally include work such as numerical experiments, computer simulations, observations or theoretical derivations.

Although we have referred to each of these components of the thesis as a chapter, you may well want to split it into two or more chapters if there is a logical division of the material.

If you begin writing your thesis on your computer as soon as you start your project, you will have less work to do at the end. Begin with your Literature Review or Materials and Methods/Experimental Techniques.

How to start writing

When you have your detailed plan, put all your notes together in the correct order and check they make sense. Include a rough estimate of the length of the different sections, keeping a balance between the components of your thesis. Show the complete thesis plan to your supervisor and make whatever changes they suggest. You can then begin writing.

Before you start writing have your plan in front of you. Use it when you write – that is what it's there for. Get hold of the correct thesis template (if available) so you type in the correct format from the start (see *Chapter 12: Layout*). Some people like to start by getting something, anything, down on paper. They then use these rough jottings as a basis from which to write their text. Others prefer to spend much longer on the initial writing, getting it as near perfect as they can. Choose whichever approach suits you best. If you are using the 'put anything down and change later' method, do not be too free in your approach to the first draft; do not go for stream of consciousness writing – it can be a nightmare trying to sort out your ideas later.

1.3 Overview of this Book

Getting organised (see Chapter 2)

Notebooks

From the start, organise your information and keep it safe. Some of your information, such as notes from experiments, will be irreplaceable, and some will take a lot of time to find again. Keep a small notebook or pad for jotting down ideas that could come to you at any time, in bed, on the bus or while eating breakfast. Some of us get our best ideas in the shower ...

Libraries

You will have access to a library at your university or college. All libraries offer the same basic services: providing books and journals, access to electronic resources such as online journals and e-books, ordering articles for you, and allowing you to carry out reference database searches. They can also supply you with other types of information. Get to know your librarian, who will be able to show you how the library and its facilities work and point you in the right direction when you are looking for information.

Your computer

We are assuming that you will type your thesis on a computer and that you have some basic word processing skills. Learn to understand and love your computer, whether you have your own or use one in your department. Your computer not only works as a magic typewriter that allows you to shift chunks of text around and correct spelling mistakes at the click of a mouse, it also allows you to access information anywhere in the world instantaneously and to sort, store and retrieve your information with ease. Take time to learn how to work efficiently and effectively with the programs you are using (browsers, word processors, graphics, reference database programs, etc.) and peripheral devices such as scanners.

Using the internet and world wide web

The internet can provide you with a lot of helpful information, from important references to moral support from other people writing dissertations and theses. This can be very useful and will allow you to sort out lots of questions immediately as you are writing. On the other hand, the internet also provides many distractions from emails to games, so train yourself not to be distracted. If you learn to turn off your email and social media while writing, you will become *much* more efficient at writing, and your social life will not come to an end if you delay checking Facebook until you have spent a couple of hours concentrating on writing – we promise ... that's what we had to do to produce this book ...

A reminder about good housekeeping

Remember to keep copies of all your computer files. We recommend using online storage of some sort as a backup to local copies. If you have not backed up your thesis and then lose it, you will have no excuses. Also keep up-to-date *dated* printouts of all your documents just in case you have a total computer disaster and lose everything you have written. A printed copy is also very useful for referring back to when you need to check what you have already written, and is essential when it comes to editing and proofreading because you may spot things on the printed page that you won't notice on a screen.

Your most important resource is yourself

Remember, you are a human being and human beings do not operate like machines – although you might well feel like one at times. Obviously, you will have to work hard on your text but do not push yourself to the limit every day. Learn to manage your time and your tasks, alternating between boring repetitive jobs and more interesting ones. Set yourself achievable deadlines for each piece of work but do not get too upset if you run over one of your deadlines, as inevitably you will from time to time.

Your References or Bibliography (see Chapter 3)

While you are working on your project, you will have come across many useful papers and other sources of information. You will need to refer to these in your dissertation so it's best to keep all the relevant details in an appropriate form right from the start. It's always a pain to try to locate references that you only half remember or which you thought you had kept but didn't. Our advice is to start a comprehensive record of the references you use as you come across them. It will save you time in the long run. You'll probably want to keep an electronic copy of the article and you must keep a record of the bibliometric details using some kind of reference database program or a word processing file. This description generally includes: a list of the authors, year of publication, title, correct journal abbreviation, volume number and page numbers. There may be regulations as to what reference details your department or university requires in your thesis, so make sure you know these before you start adding your references.

Start to collect a library of photocopies, printouts or PDFs of important references, so that you have these to hand when you need them.

Planning and writing your Literature Review (see Chapter 4)

As mentioned above, you will have been reading lots of articles about your research topic throughout your project. It's a good idea to write up your findings in your Literature Review early in the writing process. The Literature Review is a survey of all the literature (books, journal articles, websites, etc.) that might be relevant to your project. It includes an account of recent progress in the field and sets the scene for your own contribution. You could leave the Literature Review until later, but doing it early will help remind you of the significance of your results. However, note that some fields do not have a formal Literature Review, but the information is included within the Introduction.

Planning and writing your Materials and Methods/Experimental Techniques (see Chapter 5)

This chapter of your dissertation or thesis tells the reader what you did and how you did it: it is really just a recipe section. In different fields it will be called different things, for example 'Experimental Setup', 'Experimental Techniques', 'Sampling Strategy and Methodology', 'General Procedures', 'Data Acquisition and Processing', etc.

The Materials and Methods that you present must be absolutely accurate because someone reading your thesis or dissertation should be able to repeat your work exactly. Include all the details they might need, such as the pH of solutions, the names of manufacturers of chemicals and apparatus, etc. You'd be amazed how many students following in your wake in the same lab will be reading your thesis to help them with their own projects.

Write your Materials and Methods as you go along during the project, not at the end of the project. Do not worry too much about organisation of this information while you are still carrying out the project. Writing your Materials and Methods while you are still doing them will help ensure they are accurate and you have included everything. This is a tedious section to write if you leave it all until the end, when you could well be panicking or bored of the whole thing, and could easily make mistakes and omissions.

Do not simply set out your Materials and Methods in the order in which you did the experiments. Look through your materials and methods, give each a heading and then group them according to type. Within each group, put the generally used materials and methods first, followed by the more specialised ones (see Figure 1.3). If necessary, include your methodology for statistical analyses, approximation methods and estimates of error, etc. Do not include all the materials and methods you used – just the ones that are relevant to your final project.

For theoretical work, the contents of this chapter may include general theoretical methods or analytical techniques that you have

Figure 1.3   Describe any unusual – or specialist – materials or methods.

used. If you have written computer programs that could come under Materials or Methods it may be helpful to put the code (or some screenshots) in an appendix. If you have used available computer programs and databases or websites in your research, reference them fully.

Do not confuse Materials and Methods with Results. Your Materials and Methods chapter is simply a set of instructions for the reader – like giving the recipe for baking a cake. Results are what you found out from your experiments, the data you have generated, and they generally come in a separate chapter of your thesis.

Planning and writing your Results (see Chapter 6)

The Results are the core of your thesis. You need to present them well so your examiners can see what you have achieved. In your Results chapter(s) you have to explain what experiments you carried out and what you learnt from them. Remember, you are not giving detailed protocols, which are in Materials and Methods.

The first thing you need to do is get your aims clear. Why have you been doing your research? What have you been trying to show in your experiments? Try and pose your aim as a single question or statement. You can then arrange your results to best address this aim. You may also need to slightly tweak your original aims, so that your results really do address the aim. Spend time looking through your notebooks and noting all your results, even for those experiments which went wrong. Keep these notes simple – just one sentence.

Decide on your most important results and the order in which to present them. Start with the results that are the simplest and underpin your other work. Once you have set these down and are on solid ground, move to the next result, building to support your aims. Present your results in the most logical and persuasive order, and remember that this might not be the order in which they were produced. Do not present irrelevant experiments and results simply because you have them, since they may confuse your examiners and are unnecessary.

Once you have ordered your results into coherent groupings it's worth considering how many Results chapters to have. If you have a number of markedly different key results, form a separate chapter around each of them. Generally there are no rules as to how many Results chapters your thesis should have, so divide the Results into as many chapters as are necessary to group your work into logical and easily understood sections.

If you have a number of Results chapters it often makes sense to provide a brief introduction to each one, describing your strategy and specific points relevant to that section, followed by the results themselves and then a short discussion. This will give the reader a detailed critique and an immediate understanding of each result. The wider implications of all your findings can be covered later in the Discussion chapter.

For theoretical work, the Results chapters may include new theoretical results or derivations. They may also discuss the application of general theoretical methods you have developed to specific systems or examples. Just as for more experimental projects, these chapters describe the core of your work and set out what you have achieved.

Figures, tables and appendices are often extremely helpful for summarising a large amount of experimental data. Draft these before creating them and discuss the main points of each one in your text (see *Chapter 14: Figures and Tables*).

You will probably find that as you write your Results chapter(s) you think of points that should go into your Introduction and Discussion chapters. Keep a note of these as you go along, either on your computer or in a notebook.

Planning and writing your Discussion and Conclusion (see Chapter 7)

Your Introduction and your Discussion serve complementary purposes in your thesis. The Introduction starts broad and narrows down to a specific topic, while the Discussion starts with your specific results and broadens out to investigate the implications for the wider field (see Figure 1.4). We suggest you bear this in mind while planning the Introduction and the Discussion.

The Discussion is a consideration of how well your individual experiments went and why; how well they have addressed your aims; what you have found out; and how your findings fit in with progress elsewhere in your field of research. Finally, state (and justify!) the

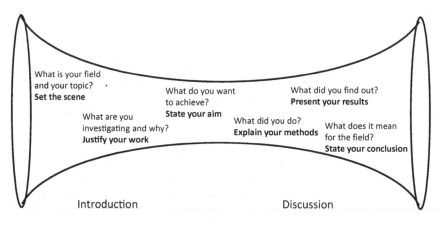

Figure 1.4 The structure of the Introduction and Discussion.

conclusions that can be derived from your work and give pointers for how the work should be extended in the future.

The beginning

Start by outlining the general thrust of your argument – restate your aims.

The middle

Discuss your results individually, carrying out any further analysis as necessary, then relate them to your field of research. Be fully aware of the background to your project, the Introduction, because this may affect your Conclusion. Remember to relate your results to current theoretical understanding of your subject. Is there agreement? If not, how can any discrepancy be resolved?

The end

Next you will give your Conclusion. This may be part of your Results chapter(s) or a separate Conclusion chapter – whichever is most appropriate for your thesis or dissertation. Make sure your conclusions are supported by your results and discussion.

At the very end, whatever your results and however successful or unsuccessful you have been, finish on a positive note by pointing out interesting avenues for future research that arise from your project. If appropriate, 'suggestions for future research' may be placed in a separate section.

Planning and writing your Introduction (see Chapter 8)

Your approach when planning the points for your Introduction should be the opposite of that for planning the Discussion. In your Introduction, start by broadly laying out the background of your research. Then narrow down to your project and the specific question you are trying to address – your aim. Finish the

Introduction with a few points about how you have tackled the question experimentally, in order to lead the reader into your Results chapter(s).

If you didn't carry out a Literature Review earlier, you must do one now, so that you fully understand the background to your research (see *Chapter 4: Planning and Writing your Literature Review*). Cite your references as you write, so you know that each of your statements can be supported by publications.

The beginning

Begin by giving the reader the background to your project. Note down all the points you want to make – just key words and simple sentences. Arrange these points so you start by describing the broad field in which you are working, then the particular area of your own research. Either give a short history of your field of study, tracing the historical development of experimental work and theoretical understanding of the area, or simply review the current situation (which is probably easier). Reference each statement with up-to-date citations. Remember that the reader cannot ask you questions while they read, so you have to provide them with all the information they need to understand your project.

The middle

Next, narrow down to your topic of interest; tell the reader why it is interesting and why you chose to study it. Cover any important experiments or theories which led to your project or which affected your work, referencing wherever necessary.

The end

State your aim and then give a brief introduction to your experimental approach in order to prepare the reader for your Materials and Methods chapter(s).

Literature Review

In some disciplines it's appropriate to combine the Literature Review with the Introduction in a single chapter. Generally, it's best to find out if there is a particular convention for this in your field. In any case, make sure that you organise your Introduction or introductory chapters into logical and easily understood sections, and that you include all the necessary introductory materials.

You may also need to include some material on the relevant theory for your project, particularly in the physical sciences. This may form a section of your Literature Review, or it may even form a separate chapter, depending upon how substantial it needs to be.

Deciding on a Title and writing your Abstract (see Chapter 9)

Composing your title

Having planned your thesis you should have a fairly good idea of what your thesis is about and should be ready to compose your title. If you are finding it difficult to think of a title, try restating your aim as a title. A short descriptive title is best, one which tells the reader about the contents of the thesis. Try not to make the title too technical.

Writing your Abstract

An Abstract is a condensed version of your whole dissertation or thesis. Your Abstract should be short, not more than one side of paper, and without headings. Make it a simple, positive and punchy account of your project that addresses the following:

- What question are you asking?
- Within which experimental system or theoretical framework are you working?
- What are your results?
- What is your answer to the question you posed?

For some degrees, such as some doctorates, you may be required to submit an Abstract and Title to your university several months before submitting your thesis. Find out if this is the case for you. Writing an Abstract is a good way to start you thinking about the contents of your thesis and give you a concise version of the final work, which will help you with later planning.

The other bits (see Chapter 10)

When your thesis or dissertation is in its final stages, add the following sections (not all of which are necessary for every dissertation or thesis):

Title Page
Dedication
Acknowledgements
Table of Contents (including Appendices)
List of Figures
List of Tables
List of Abbreviations (also known as 'Nomenclature' in some disciplines)
References (also known as 'Bibliography' in some disciplines)
Glossary
Appendices
Published Papers

Most of this material is straightforward to prepare but don't miss out any relevant items! Some word processing programs can be set up so they produce the Table of Contents and Lists of Figures and Tables automatically.

Other people's work (see Chapter 11)

Everyone who carries out research in science is building upon other people's work to some extent (remember Newton's famous statement, 'If I have seen further it is by standing on the shoulders of giants'). Therefore, when you write your thesis or dissertation, you have to get the balance right between acknowledging what other people have already done and what new progress you have made yourself. This can be tricky and it's an issue that you should discuss with your supervisor at an early stage.

Plagiarism

You must avoid plagiarism. Plagiarism is stealing other people's ideas, writings, figures or inventions and passing them off as one's own. When using other people's ideas or writings, make it clear that they are the 'property' of their author by showing that you are quoting and giving a precise reference. If a student plagiarises someone's work and the theft is discovered, which it almost certainly will be, they can be failed without further question. Plagiarism is a very big sin indeed.

Collaborative working

On the other hand, it's perfectly understandable that you may well have worked alongside colleagues during your project (especially if it is an experimental project in a large group). In writing your thesis, you will need to mention this and clarify the extent and nature of these collaborations. Take advice from your supervisor as to the best way to do this.

Layout (see Chapter 12)

The importance of good presentation

When people go for job interviews they make an effort to look clean and tidy, which shows that they are taking the interview seriously. The same is true of your thesis: if it looks well presented, your examiners are likely to take it seriously. Make sure you take into account any conventions of your discipline or department and stick to the same format and layout consistently throughout your thesis or dissertation. If your manuscript looks pleasing and comes over as a well-written and professional scientific document, the examiners are more likely to form a good initial impression of both you and your project. It's also likely to mean that the viva will be a positive experience for everyone and will focus on the scientific issues rather than clarifying or correcting unnecessary mistakes and poor style.

Thesis and dissertation templates

Some universities and departments have thesis templates available online. These give you the basic structure of the thesis (Title Page, Abstract, Introduction, Results, etc.) – all you have to do is to fill in the blanks with your own text. Make sure that you use a thesis template if there is one. If there isn't a template available, find out if your department or university has any rules about the format and layout of the thesis or dissertation. Make sure you have the correct margins and (as far as possible) the correct page layout, right from the start of your writing. This will avoid you having to make time-consuming changes to layout or the figures when you come to print the final draft. Usually you need at least a 4-cm wide left margin for binding the pages together and you will use double or one-and-a-half line spacing.

Overall content

The relative lengths of the different parts of a thesis or dissertation will vary, depending on a number of factors including the type of thesis, the scientific field and the nature of the work being described. As a very rough guide, you could expect your Introduction, your Materials and Methods/Experimental Techniques, your Results, and your Discussion and Conclusion each to be about a quarter of the total (not counting the other bits). But don't be surprised if it comes out significantly differently than these proportions. Try to avoid having an Introduction and Discussion that are massively longer than your Results section(s). If there is a word limit for your thesis, remember to use your word processing program to carry out a word count regularly so you can keep track of the length of the thesis.

Numbers, errors and statistics (see Chapter 13)

Throughout your thesis or dissertation, you will be presenting information in numerical form, whether it's specifications of equipment, results of measurements or calculations, statistical data from experiments or surveys, analysis of the results of computer simulations or parameters extracted from a fit to experimental results. Many of these pieces of information need to be presented together with an uncertainty (or error) which tells the reader how reliable the value is.

In all these cases, it is vital that you present this numerical information in a clear, unambiguous and precise manner and using appropriate units and their abbreviations (see *Appendix 7: SI Units*). There are many potential problems associated with units, values and uncertainties, and in the interpretation of statistical data. Make sure you understand how to write numerical information correctly so as to avoid any risk of confusion or, worse, inaccuracies.

Figures and tables (see Chapter 14)

Figures and tables are used for presenting your data to the reader and for explanatory diagrams or images of something you are discussing in the text. Plan their contents carefully. Discuss only their most important points in your text: you do not need to discuss every detail. Each figure (including graphs) and table needs a title, a legend and annotation. Each one must also be referred to in the text. Put figures and tables as close as possible to where you discuss them. Prepare the figures and write the figure legends for each chapter as you go along: making them can be a welcome relief from typing, and you do not want to have to do them all at the end when you will be running short of time.

Proofreading, binding and submission (see Chapter 15)

Your text will inevitably go through a number of drafts – pruning and clarifying the language each time – before it's ready for submission. With a large text such as a thesis or a dissertation you will probably go through this drafting process for each of your chapters or sections individually – some will need more redrafting than others. Avoid the temptation to go through multiple drafts with only minor changes, as this can waste a lot of your valuable time. When you complete a draft, print it out on paper – you will get a better idea of how it looks, and it's easier for most people to spot mistakes on paper than on a screen.

It is best to give yourself a break between writing and checking your drafts. It's a lot easier to spot mistakes when you have put a bit of distance between you and the draft. Try and imagine yourself as another scientist to see if your text makes sense. If you can persuade a friend or colleague to check it for you this will help, as they will spot mistakes that you have missed. Make sure they know at least a bit about your subject and have a good command of English.

Give a copy of each draft chapter to your supervisor to read and correct. Make the draft as perfect as you can, including the formatted references, so your supervisor will not waste their time correcting small mistakes such as spelling errors. Take note of your supervisor's comments and corrections (see *Chapter 17: Supervision*). Go through the draft adding information you have missed, correcting mistakes and improving awkward constructions. If your supervisor highlights particular aspects of your writing style (grammar, spelling, punctuation, scientific units, etc.) that regularly cause problems, make every effort to get out of bad habits now before you have written too much more.

Before you finally submit your thesis print a complete copy, check and proofread it thoroughly one last time. We guarantee that you will find more mistakes! But you will have found most of them by now, so take a deep breath and when you're ready, commit and submit.

Some universities now only require you to submit your thesis electronically so find out early if this is the case. If you are binding your thesis or dissertation yourself, organise what you need in advance, and check the university requirements very carefully. Also ask your supervisor for advice on what your department prefers. Some theses and dissertations, such as doctoral theses, need to be professionally bound. Find out in advance where the binders are, what they charge, how long they take and what their opening hours are (see Figure 1.5).

Figure 1.5 Check the opening hours of the binders.

The viva and thereafter (see Chapter 16)

Once you have submitted your thesis, the end of your project is in sight, but you may not quite be there yet. In addition to writing your thesis or dissertation, you may have to have a viva, which is an oral examination where examiners (usually two of them in the UK) ask you questions about what you have written. All PhD programmes and many master's programmes will require a viva. But don't worry: remember that you are the expert on what you have written. With careful preparation, the viva can be an interesting and pleasant experience because it gives an opportunity to discuss the work you have carried out with people who are interested to find out more.

Supervision (see Chapter 17)

You will have been interacting with your supervisor(s) throughout your time as a student. The nature of the interaction becomes particularly intense around the time you are writing your thesis because the supervisor's role in guiding you through this critical phase is very important. In addition, you are likely to get stressed at this stage because writing a thesis or dissertation can seem daunting, especially if you are worrying about funding or looking for a job at the same time. It's a good idea to discuss with your supervisor what support you can expect from them, and you should also be aware of what other support mechanisms there are in place in case any problems arise during the writing process.

Use of English in scientific writing (see Chapter 18)

Your text is a serious scientific document and it should be prepared with the same attention to detail as would be expected of a paper submitted to *Nature* or *Science*. Reading and understanding a dissertation or thesis can be hard work for an examiner, just as writing one is for you. Therefore, your text has to be presented clearly, without ambiguity, and with sufficient introduction to be understandable. Examiners have very little time between teaching, grant writing, administrative duties, etc., and they may even have to read parts of your text on the train, in between taking the kids to ballet practice or while waiting to get the exhaust fixed on their car.

Make the job of reading your dissertation or thesis as easy as possible: the harder an examiner finds your text to read, the less well inclined they will be towards you and your project – which is Not A Good Thing. The harder you work on making your text clear and easily understandable, the less work the examiner will have to do deciphering your text and the better disposed they will be towards you and your project – which is A Good Thing. Keep your style crisp and to the point, be concise, and use plain English. Use accepted scientific terms wherever they are appropriate. Use headings, figures and tables to break up your text into easily readable sections so both you and the reader can keep track of it. In your Introduction and Discussion you can be more discursive than in your Materials and Methods, and Results, but always keep your writing concise and focused. Use appendices for information that is a necessary part of your thesis but would clutter your text if put in with it.

Avoid wordiness, vagueness, colloquialisms and contractions (*it's, lab, isn't,* etc.) and understand specialist terminology if you use it. Use the spell-checker on your word processing program; there is no excuse for spelling mistakes. Check your grammar is correct and appropriate. It's best to avoid the passive, for example, 'the scans were continued' – it can give you the right tone of detachment, but can also be very wordy; try and use active constructions wherever possible, for example, 'scanning continued' (see *Appendices 1–5*).

A final word

We cover each chapter of your thesis or dissertation in detail in later chapters of this book and provide you with practical suggestions for presentation of data. We have also included a number of useful appendices.

Remember: start from a well-thought-out detailed plan and you will succeed!

Common Mistakes

- Inadequate planning at the start of the writing process
- Illogical ordering of material
- Lack of organisation of yourself and your thesis or dissertation
- Bad housekeeping so that essential data and drafts are lost
- Failing to proofread a printed copy of the thesis or dissertation
- Poor use of English, especially sentence construction, punctuation and spelling
- Making too many assumptions about how much the reader already knows about your field.

Key Points

- Ask your supervisor to recommend a good recent dissertation or thesis in your own field as a guide to form, content and style – and ask your supervisor *why* it's good

- Read your university or departmental rules on thesis submission, *before* you begin writing
- Find out if there is a thesis or dissertation template for your department, and if so, use it
- Make a detailed plan of all your sections and chapters before you start writing; review your work in the order: Materials and Methods/ Experimental Techniques, Results, Discussion, Introduction
- Write your Literature Review and your Materials and Methods/ Experimental Techniques throughout the course of your project if possible, and assemble your References or Bibliography into appropriate computer files
- Reference all your statements with relevant, current, references
- Set realistic goals. Have your milestones regularly monitored by your supervisor. Feel okay about taking very small steps if you are stuck
- Remember that at some point you just have to stop correcting your thesis and submit it.

Chapter 2

Getting Organised

To work well you need to use all your resources efficiently. Your most important resource is yourself.

2.1 Organise Yourself

Managing your time

Draw up a reasonable timetable for writing your thesis or dissertation. The amount of time you need for writing will vary according to the length of your project, ranging from a few weeks for a BSc or MSc project, to a few months for a doctoral study. You know at what time of the day you work best, so make sure you keep this time clear for working. Pace yourself like an athlete in training. If you push yourself too hard you will burn out and go mad. If you do not push yourself at all you will not finish the dissertation or thesis on time, or perhaps ever. You need to set yourself a series of attainable goals for each work period you have scheduled. Decide what needs to be done, and plan to get it done in the time you have allowed yourself. You will, of course, often find you need more time to complete the task than you have planned, but you need a goal to aim for (see Figure 2.1).

Managing your tasks

You need to be both systematic – making sure you do not overlook any details – and strategic – planning your work so you get it done in

Figure 2.1 Manage your time sensibly.

the most efficient and therefore worry-free manner. There will be some things that are more pressing than others, so put these high on your list of priorities. Remember that you are not an automaton; you need stimulation and changes of activity, so if you have a particularly tedious or painstaking task to do it is a good idea to break it into sections and do a bit at a time in between other more interesting jobs. Preparing a figure can be a welcome break from typing or reading references.

The best way to crack a difficult point, or to work out exactly what you want to say, is not to sit at your keyboard for hours and hours trying to force an idea when you would rather go for a walk. Go for a walk. The mind is a strange and wonderful thing. It will often come up with the answer if you leave it to freewheel on its own for a while. If it has not come up with the answer by the end of your walk do not punish yourself – your mind will just jam up. Keep calm and move on to another topic. The answer might well pop into your head the next morning.

2.2 Organise Your Information

You will need to know how and where to get information, to process this information efficiently, and to store it safely. Some of your information will be irreplaceable, and some will be difficult and time-consuming to replace. Right from the very beginning of your project you need an organised regime for looking after your data, notes and references.

Pens, paper and Post-it® notes

These are very basic resources, and you should have a ready supply of them. You need to be able to jot down bits of information and ideas whenever they come to you. Post-it® notes are very useful for sticking notes in journals or books you are reading.

Notebooks for keeping ideas and for information

You have been keeping a lab book with a record of all your experiments (or the equivalent for other types of work) throughout your project. This is going to be the most important document for you to refer to when you come to writing your thesis or dissertation. It should include an account of everything you did, all the relevant information about the conditions, and what you found out. It should also provide a key to all your data (files, plots, images, etc.) that will allow you to locate the result of any piece of work.

Also keep an 'ideas' notebook for putting down odd ideas about different aspects of your project, for example, points for the Introduction or Discussion. The more you learn and think about your project, the more you will want to discuss and explore different themes. You will have stray ideas – the first inklings of connections between different bits of information. These could occur to you at any time and you might have to scribble them down on the back of bus tickets and beer mats. If so, copy them into your ideas notebook, or at least stick your bus tickets and beer mats into it as soon as possible so you do not lose them. Very often an idea that has been eluding you all day will come to you in the night. Scientists have their muses and should have something on which to write ideas when inspiration comes, so keep your ideas book (with a pen or pencil) by your bed. If you wait until the morning the idea could well have vanished into thin air – or have become so deeply mixed up with that dream about you, Justin Bieber/Pink/Brian Eno/Jackie Chan and the Gold of Deadman's Bay, that you can no longer quite work out its relevance to your work on nuclear fusion. Some people prefer to store their ideas as they crop up, by leaving voice memos on their mobile phones.

You can use a different section of the same book for noting the main points from papers and articles you do not wish to print, or for information that you come across in seminars or posters at scientific meetings.

Use a notebook which has the advantage of being bound so you are unlikely to lose pages from it, or use a ring-binder, which has the advantage of being easily organised and reorganised as you go along. If you are using a notebook, it can be useful to leave three or four pages blank at the beginning for a contents page. Of course you may prefer to keep entirely electronic records, which is fine, but sometimes a paper notebook is easier for jotting down ideas.

Keep your ideas notebook and any other books well organised and safe, and if possible in a place where they will not be disturbed. Do not leave them lying around in pubs or on trains.

Making notes

Make your notes easy to follow. There are many different ways of taking notes – use whichever you find most comfortable. Use different coloured pens for different themes if it helps. Putting individual concepts or items of data into bullet points is a good way of summarising simple ideas. Flow diagrams help show sequences of events and connections between ideas. However you take notes, always put them in your own words and try not to write down too many details – if you are taking notes from a reference you can always go back and re-read the paper. Pick out the main points of interest, and summarise them. Never simply copy the ideas word for word, or you will be in danger of unconsciously plagiarising – it is very easy to copy your notes straight into your text forgetting that the words are someone else's. Make sure you know exactly where your information came from – write down either the date and occasion, or the reference (see *Chapter* 3: *Your References or Bibliography*).

Some people like to use highlighting pens or underlining to pick out important points in printed copies of their references. Use this system if it is helpful, but do not highlight every detail and take care not to plagiarise when writing your text from these notes.

Increasingly people now keep notes in an electronic format, by highlighting and annotating Word documents or PDF files of papers. One advantage of this, apart from cutting down on the use of paper, is that you can then take all your notes with you on a laptop wherever you are, without having to go back to your desk and rummage through the drawers looking for a copy of an article that you have scribbled notes on. If you work in this way, just make sure that you have a reliable backup of all your files in case of disaster like a stolen laptop or a crashed hard drive.

2.3 Sources of Information

Textbooks, dictionaries, English-usage books and thesauruses

Keep to hand any regularly used textbooks or reference books. Buy a dictionary, because you need to know the meaning of words you are using. As well as knowing what the words mean, you need to know how to use them, so it is a good idea to buy yourself an English-usage book. A classic, which professional writers use, is Fowler's *Modern English Usage* (Oxford University Press). This is the recognised authority on conventions in current-day English, and is also an interesting reference book to browse through.

A thesaurus can be very useful if you find yourself using the same word over and over again and feel the need to vary your prose a bit. It is bad writing style to use the same word more than once in a sentence and several times in a paragraph; it is jarring to read this kind of English. The word *thesaurus* comes from the Greek word meaning store, treasure or storehouse. It is just that: a treasure house of words arranged into groups with similar meanings. If you are stuck for a word, or have one that you know is not quite right, you can use a thesaurus to track down a suitable alternative.

Libraries

All libraries offer the same basic services, such as providing physical copies of textbooks and journals, giving access to electronic books and journals, and ordering articles or books for you from other libraries. They have facilities that allow you to carry out reference database searches, and have information about what resources are available from other libraries. They also hold useful specialised reference books containing information such as citation abbreviations, compendiums of physical and chemical data, and general reference books, for example encyclopaedias, dictionaries and thesauruses.

One of the most useful resources you will find in the library is the librarian. They know how the library works and will be able to either find information for you, or show you how to find it for yourself. If you have only a vague idea that some information might exist, or just want to know what information might be useful to you, do not hesitate to ask the librarian. The more focused you are in your question, the easier they will find it to help you.

The internet

While carrying out your research project or writing your thesis, you will probably have used either information or programs from many different websites. You need to reference these sites by giving their name and address (http://www. ...) so that any examiner or fellow scientist can see exactly what you did. You could put all the addresses into a table, the References section or in footnotes at the bottom of the page. Always give the date you accessed the site as well as the address because content can change with time.

Online searches

You need to be aware of the literature (books and journal articles, for example) that might be relevant to your project. Use a computer to carry out a search of the literature databases (such as Web of Knowledge, PubMed, Google Scholar, BIDS, MEDLINE, SPIRES, Compendex, Inspec, or Chemical/Physics/Biological Abstracts) with keywords from your subject area, and also browse the internet using key words. It is up to you how many years back you go. If your keywords have brought up huge numbers of references, then start by selecting only those most obviously relevant to your project, and the most recent. Be imaginative in your use of keywords, and try searching with common abbreviations as well as complete words and phrases, and alternative spellings if they are relevant.

When you find useful references, download the reference details directly into your own computer database. Read the references, not just the online abstracts (see *Chapter 3: Your References or Bibliography*).

Most of your references will come from books and journals or other sources of published information including other people's theses or dissertations, newspaper articles, conference proceedings and catalogues. You can also cite information from websites, lectures, talks or unpublished works if you have to, but it is best to stick to published material so the examiners can read your sources. If a work is not in the public domain – for example, if it is in a private collection, a private conversation or a technical report that is only available on request – ask permission from the owner/author before citing it.

Building your own library of papers

During the course of your project you will come across papers in journals and other publications that you will want to use when you come to writing your dissertation or thesis, particularly for your Introduction, Literature Review and Discussion. It is a good idea to keep an electronic copy or take a photocopy of any papers you think will be useful to you. If you have your own copies, you will be sure of having them available when you need to use them.

Information from commercial catalogues

Suppliers of equipment and materials, like chemicals or molecular biological resources, provide a lot of information in their catalogues, which could be useful when you are writing your Materials and Methods. Use the catalogues as another of your information sources, referencing if necessary of course.

Information from microfiche

Some older types of information may be stored in a very small font on a type of transparent plastic card called a microfiche. To be able to see the microfiche you need to put it in a microfiche reader, which can also produce a printed copy for you to take away. Most large libraries either still have microfiche readers, or know where to find one. If you are using anything from microfiche, remember to reference it.

Information from CD-ROMs and DVDs

Many important reference works are available on a CD-ROM or a DVD. If you use these resources, don't forget to reference them properly.

2.4 Copyright

Copyright affects what material you can copy and keep for your studies and it also arises in the context of what you can reproduce in your thesis or dissertation – we discuss this in *Chapter 11: Other People's Work*.

Copyright protects the intellectual property of an author in whatever form it is recorded: books, journals, photographs, compilations, offline databases, computer programs, CD-ROMs, DVDs, etc. In the UK and the USA, for example, copyright lasts for 70 years after the death of the author. Copyright means that no one can reproduce an author's work without having first gained permission. For private study, research and criticism you are allowed to reproduce sections of someone else's work under a convention known as 'fair dealing' ('fair use' in the USA). Very briefly, fair dealing allows you to photocopy other people's work to a limited extent, for private study. **Check with your library if you are at all unsure about copyright restrictions.**

2.5 Your Friendly Computer

Your computer is your friend. It is an essential everyday piece of equipment. From the start, store as much of your information as possible on your computer, as this will save you time when you come to write your dissertation or thesis at the end of your project. When you begin the writing proper you will be working most of the time on a computer, word processing your information. If you can't always use the same computer, make sure you can access your files wherever you are, either by taking a memory stick with you or using an online storage location.

Guidelines for computer usage

In most work environments there are strict regulations governing how and for how long people should work with a VDU (Visual Display Unit) or computer monitor without a break. The regulations are there for a good reason. People are not machines; they need breaks, they need cups of coffee, they need to go to the toilet, to get some fresh air, or just a change of scene from time to time.

Make sure your chair is comfortable (see Figure 2.2). You need the lumbar region of your back to be well supported. The chair should be set to the right height, so that your thighs are horizontal and your lower legs and torso are vertical. The keyboard should be at such a height that your arms fall comfortably to your side, with your forearms resting horizontally. Keep your wrists as straight as possible and avoid having to twist them. Ensure there is enough space between you and the screen so that you do not have to squint to get it into focus. It's generally easier to find a comfortable position with a PC than with a laptop, so if you are using a laptop most of the time, you should take particular care not to get uncomfortable.

The image on the screen should be stable with no flickering. Adjust the brightness and contrast so you can easily see what you are

Figure 2.2   Are you sitting comfortably?

typing. Also make sure that the lighting in the room gives a good contrast between what is on the screen and the surrounding area. Most importantly, there should be no glare from electric or natural light reflecting from the screen, which will tire your eyes quickly.

Take regular breaks. It is best to take a number of shorter breaks frequently rather than waiting until you are tired and then taking a long break in order to recuperate. A break of about five or ten minutes every 50 or 60 minutes is recommended. Take these breaks away from the screen: go for a run, have a cup of tea, walk the dog, bake a cake; do something that does not involve using your computer. You will be able to find more detailed information in the 'Health and Safety at Work' section in your library.

Saving your information

Remember that computers can sometimes crash without warning; whenever you stop to gather your thoughts for a second, or come to the end of a section, use the Save command. Get into a habit of saving regularly so that it becomes automatic. There is nothing more frustrating than stretching back with a satisfied yawn after three hours typing, thinking what a good evening's work you have done, to see it all disappear forever as your computer crashes and wipes all your unsaved work.

It is best to work on files that are on the hard disk and to keep the master copy of all your work there. Regularly copy from the hard disk to a memory stick or some other backup device while you are working on your document. Probably the safest place is a server that is itself regularly backed up. This is especially important if you use a laptop. If the one and only copy of your text becomes corrupted, you will have lost everything. Keep your backups safely stored in different locations in case one is lost or stolen. Or email your latest versions to yourself so that they are stored remotely. There is no excuse for losing files!

Organising your work in folders or directories

Make sure you have a sensible archiving system for your computer files. This will help you see what files you have and which draft you are working on. It will also help you to avoid losing valuable files, which you might delete by mistake. Organise your files into different folders or directories, and name the folders or directories so that their contents are obvious.

As well as naming your files (and folders or directories), get into the habit of dating everything, either within the text itself (you could put this information in the header or footer zones) or in the file name. Remember to change the date when you create a new version so you do not risk muddling multiple versions of the same file. It is a good idea to have a 'current' folder or directory for your most recent work, and 'old' folder or directory for older drafts, which you still want to hang on to. Do not be afraid to get rid of information once it is no longer any use to you, so clear out your 'old' folder or directory periodically, or it will just get cluttered with redundant information.

Typing

We are assuming you will be typing your dissertation or thesis yourself. How well you can type will affect the speed and ease with which you produce your text. Two-fingered typing will get you through, but the more fingers you can use and the more accurately you can use them the easier the task will be. It will save you time in the long run if you learn how to type to a reasonable standard before starting to write your text. You could go on a typing course, or use one of the typing tutor programs available for use with most computers. You could ask the librarian in your institution to recommend a good one.

Learn to love your computer

However new your computer is, it will crash from time to time and your programs will develop glitches. You need to learn at least the basics of how your computer and programs work so that you can deal with these problems when they come up. We are not advising that you become a computer engineer on top of all your other studies, but if you make the effort to familiarise yourself with your equipment and learn to recognise easily fixed problems, you will save a lot of time that would otherwise be spent waiting for someone else to sort out the computer. Most of the time the trouble is fairly minor. Most online manuals have sections entitled 'Troubleshooting' (or something similar) that can help you fix the common glitches. You could also ask your university's computer support staff for advice. If your computer develops a major glitch then you have no choice but to call in an expert, immediately.

A final comment about working on your computer: don't let yourself get distracted by other things when you are meant to be writing your thesis. Facebook and emails are great things but the distraction they cause will reduce the efficiency of the time you spend at the computer enormously. So *turn off everything else* when you are writing (this goes for your mobile phone as well of course). Your thesis is important and for once other things will have to wait till you take a break.

2.6 Word Processing

Fortunately for the computer-phobic, word processing programs are produced for everyone to use, from sous chefs to skydivers, from ecologists to molecular modellers, so they are accessible for beginners. We have assumed that you have some basic knowledge of word processing. How much you learn about your word processing program is up to you, but learning a few commands, other than simply how to type and cut and paste will save you a lot of time when you write your thesis or dissertation and will help you produce a much more professional-looking manuscript. Time spent finding out about shortcuts is probably a good investment in the long term.

Nowadays most theses are written using Word (or the open source equivalent, OpenOffice). In the areas of the physical sciences and mathematics, LaTeX is commonly used as an alternative because of its excellent equation-handling capability.

All word processing programs are slightly different, so we will not go into the details of the commands. The best advice is to read the online instruction manual that comes with your word processing program, to use the 'Help' option in the program, and to play around with the program, experimenting with the different commands. Learn how to find any specialised symbols you need, such as mathematical symbols, Greek letters, and accents. Learn about commands using combinations of keys rather than working with the mouse alone, because key commands are often faster. Make sure you know how to use tools such as the spell-checker, how to find and replace, and what commands to use for layout (see *Chapter 12: Layout*).

Giving one simple command to write something long and complicated

If you have a long complicated word or phrase you use frequently, it is a nuisance to have to type it in full every time you need it. You can save yourself time and finger-work by making use of macros, or short cuts, which enable you to have the computer add a particular word or phrase to the text every time you give a simple command. For example, you can arrange things so that every time you need to write *Saccharomyces cerevisiae, scanning tunnelling electron microscopy,* or *electromagnetic resonance imaging,* you just press the control key followed by a key of your choice. Have a look at the 'Help' section of your word processing program.

Using a spell-checker

All word processing programs have spell-checkers and so it is inexcusable to make spelling mistakes in a thesis or dissertation. When you are setting your spell-checker, be aware of the differences between American and British English, most of which are to do with spelling. Set your spell-checker appropriately: if you are studying in Britain it is best to use British English. But don't get lazy with spelling – remember that a spell-checker will only pick up mistakes that don't make another word.

Using a thesaurus

Many word processing programs have a built-in thesaurus. If yours has one make use of it when you find yourself using the same word over and over again.

Using a grammar-checker

It can be interesting and instructive to use a grammar-checker, but it may not be worth using one to check your whole text: it may not correctly recognise the structure of complex sentences. Grammar-checkers pick up only some, not all, of the likely mistakes, for example:

> While undergoing treatment, the doctors found that their patients' recovery rate was improved by listening to classical music . . .

This is grammatically correct but is probably not what the writer would wish to say. It means that the doctors were undergoing treatment when they made their finding (see the section on dangling participles in *Chapter 18: The Use of English in Scientific Writing*).

Number chapters and sections automatically

Most word processors have a command that automatically numbers your sections (see *Chapter 12: Layout*). If you move the headings around then they will automatically be renumbered. This can be useful if you have your notes on the computer and want to rearrange them as your planning progresses. These commands are different for each program, so read the manual or online help.

Create a Table of Contents

Many programs have an extremely useful command that not only numbers your headings, but remembers them, notes which page number the headings are on, and automatically creates a Table of Contents for you. This command can save you an enormous amount of time when you come to finish your thesis, and is worth finding out about right now.

Word processing with mathematical symbols

All word processors allow you to put in simple mathematical symbols and write equations, but most programs have been designed for use with everyday English, and so typing the symbols using a normal word processing program can take forever. Therefore many scientists work with word processing programs or add-ins that have been specifically designed for easily and speedily incorporating mathematical symbols and formulae. If you are in any doubt about which program to use, talk to your supervisor or ask your computer department for help. We strongly recommend that you use one of these programs if you are writing a project with a large mathematical component. Or swap to LATEX, which deals elegantly with all types of mathematical expressions. You may find it takes a while to get the hang of this package, but for a heavily mathematical thesis it's worth the effort.

Thesis templates

Some universities and colleges have what they call a 'template' for theses and dissertations, which will be available on a home page. Institutions provide templates to ensure their students' dissertations and theses conform to a standard layout. If using a template, no one has any excuse for missing out sections of their text or important details such as the information required on the Title Page. Find out if there is a template available for you and if there is, use it.

2.7 Other Useful Software Packages

Reference database programs

In the pre-historic pre-computerised days of thesis writing, every time we cited a paper in the text we had to write the citation in full and then type the full reference details at the end of our theses. It was extremely painful. Typing long citations many times (for example, *Forster, Romero, Albertine and Pollitt, 2014; Mortimer, Hendy and Coldwell, 2015; Borrell, Sullivan Kaplan, Stitz and Robertson, 2016; Rowbottom, Carr, Brown, Hewitt and Cieka, 2016*) was extremely boring and led to many typing mistakes. It was also easy to forget to add or remove references from our reference list when we edited citations in our text. Reference database programs overcome these problems (see *Chapter 3: Your References or Bibliography*).

With a reference database program, all you have to do is type in the full reference details of a paper once or (more usually) simply import them from the internet. If you automatically import reference details from reference databases, such as Web of Knowledge, BIDS or MEDLINE, this reduces the chance of typographical errors occurring and cuts down on the amount of typing you have to do.

The reference is put into an electronic library on the computer and is given its own unique tag. When you cite the reference in the text, you just copy and paste the tag (which can include the name, date and a symbol) – no more typing out names or dates or numbers. When you come to print the thesis, the program automatically puts the citations and full reference details into any format you like, only giving details of the citations you have placed in the text.

When you have formatted your references and are ready to print a draft, first check that all the references have in fact been properly formatted. If they have not been, the tag will not have been replaced by the appropriate citation. Run a 'Find' for the tag symbols to detect any stray ones.

Most reference database programs are very easy to use and are invaluable to anyone citing a lot of references. Learn to use a reference database program immediately. If you go on to further study you can build up a large library of such references over the years.

Spreadsheet programs

These are ideal for storing large amounts of data suitable for placing in a table or figure. You can insert your data into a spreadsheet, while you are gathering them, and prepare your tables and figures directly. You can then import these from the spreadsheet directly into your text when you come to write your thesis. This will save you time, and cuts down on the chance of making a mistake while retyping the data (see *Chapter 14: Figures and Tables*).

In many cases, graphs of your results produced from a spreadsheet program will be sufficient for your purposes and you will not need to make use of more complex and powerful packages. As a general rule, don't bother to get proficient in a specialist package if you can get the results you need from a program you already know how to use.

Data analysis and graphing programs

There are many specialist graphing programs around that will automatically create graphs from data that you enter. These programs can be fairly simple to use, at least for basic tasks, and if used well will produce beautifully drawn curves and lines, saving you a lot of time with graph paper and rulers. Usually, the programs work by having a spreadsheet into which you place your data, then you have a choice of graphs that the computer can create from these data. For example, you could produce a histogram, or a pie chart, or a scatter graph, or a curve using points either with or without error bars. You will also be able to add fitted curves to the plot if necessary. Usually your data are only appropriate for one type of graph, so the decision is easy, but if in doubt go to your supervisor for guidance.

Remember to fully annotate your graph when using these programs, for example, state what the x- and y-axes represent and clearly indicate a selection of values along these axes.

Be wary of going over the top using the facilities that are available to you on these programs. Seminar speakers quite often make the mistake of presenting amazing multi-coloured, unnecessarily three-dimensional graphs, in which the actual data are lost in a mass of fancy graphics. The simplest graphs are usually the most striking and effective way of presenting your data (see *Chapter 14: Figures and Tables*).

Graphics and drawing programs

Like all other computer programs, graphics and drawing programs come in various shades of complexity. Small children play with simple graphics programs, producing fat ellipses and large squares and telling people it is a picture of Mummy and Daddy. At the other end of the expertise scale, professional graphics and video design artists produce very complex graphics with similar programs. Most of us lie somewhere in between these two extremes.

We are going to generalise wildly about these programs and really you need to find out for yourself what your own resources are: graphics programs allow you to produce complex images from relatively simple components such as circles and squares and straight lines, with or without colour. They are most useful for producing line drawings and diagrams, and for simple annotation of previously scanned images. Drawing programs tend to be used more by professional artists and are often a lot more flexible but less user-friendly. Many scanners use drawing programs for altering and manipulating scanned images, for example, to reduce the intensity of the background or increase the intensity of the foreground.

The trick to working with all graphics and drawing programs is to get hold of the manual or read the online help, and, most importantly, play with the programs. We encourage our students to design their party invitations, play around making Christmas cards or send fancy letter headings to their friends using the graphics and drawing programs that we have, because this is an excellent way to become familiar and at ease with the process (see *Chapter 14: Figures and Tables*).

Common Mistakes

- Running out of time for writing up before a deadline
- Disorganised notes
- Losing work through careless management of files
- Failure to use a spell-checker correctly
- Working in uncomfortable conditions: spending long hours in a bad posture in front of the computer will lead to backache and eye strain.

Not all computer problems are soluble. A former colleague who is now a senior professor working in UK, and an FRS (look it up), did his PhD in a department that was one of the first to get a large mainframe computer. Working late in the lab one night, he accidentally knocked a litre of boric acid down into the back of the computer. Afraid to admit to the disaster he watched, terrified, as this expensive new computer developed more and more faults over the next few days. He developed bad eczema as his situation – and that of the computer – worsened. Finally, someone took the back off the computer and discovered the rotting heap of wires inside. The graduate student owned up, and his eczema improved, but not his supervisor's temper.

Key Points

- Set a realistic timetable for achieving realistic goals
- Make full use of your library
- Put your references into a reference database program immediately

- Take care NEVER to plagiarise when copying from notes or references
- Use a thesis template if it is available for you
- Save all computer files on a hard disk and a memory stick or server
- Learn to use your word processing program beyond the basic commands, and make full use of other programs such as reference database, graphics and graphing programs
- Notebooks are an invaluable and often irreplaceable source of information, so make an effort to keep them safe, and write some sort of contact details in the book in case it gets lost.

As an undergraduate a colleague of ours had a bag containing an entire year's worth of notes stolen from a car a week before his exams. It was not a happy experience, particularly as all the notes had been borrowed from a friend.

Some years later, one of us was supervising student who spent a considerable amount of time (about six months) collecting data, storing and analysing it on a laptop, and carefully backing up the laptop onto disks. Unfortunately the computer AND the backup disks were stolen when the bag they were all kept in was removed from the back of a car ... and no, there was no other backup ...

Chapter 3

Your References or Bibliography

3.1 References and Citations

To avoid confusion, we use *to cite* to mean to refer to a piece of work in your text, and *citation* to mean the indication in the text that you are doing so. Here are two types of citation:

> The plumage on most southern African landfish is yellow (**9**), but specimens with blue or red plumage are seen occasionally (**10, 11**), similar to that found on Egyptian landfish (**12–14**).

> The plumage on most southern African landfish is yellow (**Martin *et al.*, 2015**), but specimens with blue or red plumage are seen occasionally (**Collins *et al.*, 2013, Gabriel, 2014**), similar to that found on Egyptian landfish (**Michael and Ridgeley, 2015, Hoffs *et al.*, 2016, O'Riordan *et al.*, 2017**).

By *reference* we mean either the original document or the directions for the reader to find the original document, which will be given in your 'References' section or 'Bibliography':

9. Martin, C., Berryman, G., Champion, W., Buckland, J. (2015). All yellow predominates in southern African landfish colouration. Landfish Today 28: 5–23.
10. Collins, P., Banks, A., Rutherford, M. (2013). King crimson landfish. Landfish Genome 13: 5–23.

11. Gabriel, P. (2014). Landfish without frontiers. Landfish Newsletter 52: 5–23.
12. Michael, G. and Ridgeley, A. (2015). Hertfordshire and Mediterranean landfish plummage. Landfish Today 27: 102–123.
13. Hoffs, S., Peterson, D., Peterson, V. (2016) Egyptian landfish plummage and perambulation. Landfish Newsletter 54: 305–348.
14. O'Riordan, D., Hogan, N., Hogan, M., Lawler, F. (2017). Animal instincts and colouration in landfish. Journal of Landfish 109: 75–94.

3.2 Using References

Any piece of scientific writing, whether it is a journal article that presents important new results or a first-year essay, has to be properly referenced because the reader must be able to check the original sources of any statements you have made.

What to reference

Strictly speaking, you should reference any information you have used that is in the public domain, i.e. that can be found and read by your readers. This includes other dissertations and theses, your own published work, and conference proceedings. In practice you do not need to reference statements of general knowledge ('the earth is round') but you do need to reference information that is not yet generally accepted or known ('core samples from the University of West Cheam contain extra-terrestrial rocks' (Bowie *et al.,* 2015)). If you have used any information from your own publications, remember to reference these as well. You should also reference methods you have used, including computer programs.

What to cite

Give a citation in your text every time you make use of a reference. You can cite information for which no written reference exists, for example a conversation or seminar talk, but avoid this if possible because the reader has no way of checking your information. If you do this, check the statement with the speaker and add 'personal communication' or 'pers. comm.' to make it clear where the information comes from, for example,

> The University of West Cheam is built on earth from the Valles Marineris region (Professor Robert Smith, personal communication).

If you cite work that is not in the public domain, make sure you have appropriate permission before doing so.

How many references and how old?

There are no rules as to how many references a thesis or dissertation should contain, as long as the reader can see where your information comes from. A first-year report may cite only 20 or so references, while at the upper end of the scale a well-referenced life sciences PhD thesis can cite 200 or 300 papers. If you are unsure whether you have referenced your statements sufficiently, ask your supervisor to check.

Science is very fast moving and you should ensure that your references are up to date. If your most recent reference is five years old this is a clear sign to the examiners that you have not been keeping up with your field and are unlikely to have a good background knowledge, or even to be particularly interested in your research. Make a point of checking that you have included the most recent relevant references.

Problem references

Sometimes it is impossibly difficult to track down an original document, perhaps because it is very old or in an extremely obscure journal. In this case, indicate you have not read the original but know that it is useful because you have seen it cited in other papers, by writing *cited by* ... or *cited in* ... or *reviewed in* ... However, be very cautious of using other people's references in this way as you are putting a lot of faith in their judgement and thoroughness.

Quoting

If you need to quote the exact words of an original statement, make sure you get them right. If you are using short quotations, less than about 20 words or so, you can simply put them in inverted commas in your main text (see the punctuation section in *Chapter 18: The Use of English in Scientific Writing*):

> The University of West Cheam is known for its unusual microhabitat and Professor M.L. Ciccone stated 'local flora include the rare Tharsis Montes hydrangea', which is a food source for northern landfish.

With larger sections of text it is better to put quotes in a 'block indent', with the citation following, like this:

> Heat generation from normal bacterial metabolism remains a difficulty for researchers in this field. Many attempts have been made to overcome this problem, Karloff maintains:
>
> > The normal phases of catabolism, in which complex substances are decomposed into simple ones, and anabolism, in which complex substances are built up from simple ones, have been bypassed by the technique of Moog Genesis (Karloff, 1959).
>
> These claims are highly suspect and many other studies show that such measures of energy function may be artefactual and arise from the unusual measurements taken by Dr Karloff at the University of West Cheam.

Copyright

There are strict rules about photocopying and quoting other people's work (see *Chapter 11: Other People's Work*).

3.3 Collect and Store Your References from Day One of Your Project

Build your own electronic library – use a reference database program

When you come across a useful reference, you need to keep a computer record of the reference details: the names of authors, year published, title, journal name, volume, page numbers. Keep your references in a reference database program because this is the best and easiest way of storing and retrieving this information. All universities and colleges use these programs, so talk with your supervisor about them.

If possible, keep PDF copies of all the key references too, so you can easily refer to them again when you need to (and print them out as necessary). In this way you will gradually build a personal library of the original papers you are using as references, so you will have them to hand when you need them (see Figure 3.1).

Start now! YES RIGHT NOW

Your references are the single most tedious item you will have to type into your thesis. If, from the start of your project, you enter references into a database as you come across them, you will be able to insert the citation electronically into your text as you write, and you will not interrupt your flow of thought. Your reference database program can probably import references automatically from elsewhere, so you don't make horrible mistakes in mistyping people's nomes, for example ...

If you have not been using a database from day one, even for a relatively short thesis or dissertation, then start entering your references into a database RIGHT NOW! It will get you back in contact with parts of research you might have forgotten, and will save you having to do this painstaking task during the thesis writing process when your time could be much better spent honing the more interesting and creative aspects of your writing. If you do not do this, you will have to stop and track down the correct papers every time you need to cite a reference, then enter them into the database, then enter the citations into the text. As we said earlier, most reference programs allow you to import all the relevant details of each reference directly (e.g. from a website such as Web of Knowledge), so you don't need to enter everything by hand.

The absolute worst-case scenario is that you do not use a database program at all, and try citing references and entering them into the References or Bibliography section by hand – this *always* leads to mistakes: missing references, missing citations, incorrect references, etc. The longer the thesis, the more the mistakes. Use a database program. Really.

Figure 3.1　Build your own library of references.

3.4 The Final Text
Add citations to your text as you write

When you are writing the text of your thesis or dissertation, and you make a statement that needs referencing, put the citation in your text as you go along rather than writing your text and adding citations later; that way you will be able to make sure any information you are presenting has a reference that goes with it. Citations are usually placed at the end of the sentence so that the flow of the sentence is maintained.

Statements or methods that are presented in theses without being referenced imply the writer may not have a good background understanding of their field, or, far worse, made it up as they went along. If you cannot find an appropriate reference to cite it may mean that you need to do some more reading around the subject and explore a little more deeply before you continue writing. Choose your references with care, and make sure you really have read them; by citing a paper you are saying 'I have read this paper'. Your examiners may well be interested in the content of the paper and it doesn't give a good impression if you cannot answer questions about it.

The final printout

When you have the final version of your thesis, run your eyes over the Bibliography, and just check that all your references are up to date, and fully listed with their correct authors, title, journal, volume number and pages. Also check the Bibliography has a consistent format and that if you've used an abbreviation for a journal name it is the same abbreviation throughout – and, because journals have standard abbreviations, that it is the correct abbreviation! Standard abbreviations can be found on journal websites or on databases such as Web of Knowledge (http://wok.mimas.ac.uk) or PubMed (http://www.ncbi.nlm.nih.gov/pubmed).

3.5 Formats for Citations and References

Scientific journals have strict rules as to how papers are presented, including how the references are cited. Check if there are any regulations concerning the format of your references. Some universities and departments require references for theses to be organised according to the Harvard System of Referencing and others according to the numerical system (see below). If there are no specific rules for your thesis or dissertation it is worth finding out if there are any conventions in your field or in your department. If there are, you would be sensible to follow them – ask your supervisor and have a look at a good, recent thesis in your field.

The Harvard System of Referencing

With the Harvard System, the names of the authors and the date of publication are included in brackets in the citation in the text:

> ... Coulomb explosion imaging is of limited use in this field and so was not used in building the Tharsis tunnel (Carter and Knowles, 2016).

This system can get a little unwieldy, and it is best to avoid citing too many references in one block, but it reminds both you and the reader which papers are being referred to, which can be very helpful in an oral examination if you are asked to go through aspects of your thesis. Where the name of the author occurs naturally in the text you only have to add the date in brackets:

> ... thus Björk's assertion (2014) that partons, such as quarks, are actually energised fairies is extremely controversial.

> ... as Rowlands and Simons (2014) state, the β-form of cheamium sulphate is made up of molecules in an hexagonal array ...

If there is more than one author, cite both names on two-author papers. For papers with three or more authors, cite the first name only, followed by the words *et al.*, from the Latin, *et alii* meaning 'and others'. Normally *et al.* is italicised because it is a Latin phrase (see *Chapter 18: The Use of English in Scientific Writing* and *Appendix 5: Latin Words and Abbreviations*),

... if a pair of opposed and external forces is applied to a body, the relative positions of this body's components change (Jagger, 2011, Watts *et al.,* 2012, Richards and Wyman, 2013), and new restoring forces arise (Reid *et al.,* 2015, Glover and Calhoun, 2016, Wimbish, 2017).

You will often have to refer to more than one paper, in which case it is usual to cite them according to the order in which they were published, starting with the oldest paper (Adamson, 2013, Watson *et al.,* 2014, Brzezicki and Butler, 2015). When you cite papers from the same year, put them in alphabetical order according to the name of the first authors.

Whilst you would not normally include initials it makes sense to do so if you are citing two or more papers with the same surname and date (for example, Hawkins, D., 2014, Hawkins, J., 2014). When citing two or more papers with the same first author in the same year, indicate this by adding a, b, c, etc., after the name or year, for example, Jackson, M. (2012a); Jackson, M. (2012b).

With the Harvard System, the references are arranged in alphabetical order by the surname of the first author of each paper, in your References or Bibliography section at the end of your thesis. If you have many papers with the same first author, arrange them with the earliest first, and most recent last.

Numerical system

Many institutions and disciplines prefer the numerical system:

> ... if a pair of opposed and external forces is applied to a body, the relative positions of this body's components change (5–7), and new restoring forces arise (8, 9).

This does not disrupt the flow of the writing as much as the Harvard System, but the reader has to look in the References section to see what the number relates to.

If you are using the numerical system you can put your citations in brackets:

> Students at West Cheam University have undertaken interesting research that shows the time-space continuum is distorted along

the axis of a ley line running between Nonsuch Park and Uppingham (9). Dr Karloff, Dean of Alternative Physics at the university, disappeared for 30 minutes during a seminar while demonstrating the effect of Moog Genesis in the vicinity of this ley line (10, 11).

or some disciplines prefer to put citation numbers in superscript:

A recent study[12] has revealed disturbing evidence of fluctuation among time zones in South Kensington and Bloomsbury.

Generally when using the numerical system the reference details are arranged in numerical order in your References section; in a few subjects it is acceptable to place the reference in a numbered footnote at the bottom of the page containing the citation.

3.6 How to Write the Full-Length Reference

There is a minimum amount of information that you should give to ensure the reader can find the original reference. For example, for journal articles you need to include: the names of the authors, the name of the publication, the volume number and year, and the first page number of the article or chapter (or the electronic locator such as the digital object identifier, or d.o.i.). Try to make things easy for the reader by including as much information as possible. In particular, it is helpful to include the title of the article.

References from journals

Here are a few examples of possible layouts – which do you find easiest to read? The article is by the authors Kravitz and Winwood, it is published in 2016 in the journal *Landfish Today*, the title of the article is 'The lifecycle of landfish in east coast resort towns', and this article appears in volume 29, pages 8–12.

Kravitz L., Winwood S. 2016. The life cycle of landfish in east coast resort towns. Landfish Today **29**: 8–12.

Kravitz, L. and Winwood, S. The life cycle of landfish in east coast resort towns. Landf. Tod. 29, 8–12 (2016)

Kravitz L; Winwood S (2016). The life cycle of landfish in east coast resort towns. *Landf Tod*. 29, 8–12.

Note that scientific journals all have their particular requirements for the format of references. Many of these are only slightly different (for instance, they may require a comma or a semicolon separating authors' names; there may be an additional 'and' or '&' before the name of the final author; and various parts of the reference may have to be in *italic* or **bold** font). If you are submitting an article to a journal you have to be careful to follow their format exactly; however, for your thesis you can generally choose whichever format you prefer.

When referencing journal articles, make sure you include the following:

Surname of all the authors and initials

It is best to cite the authors in full unless this would be very unwieldy (if for example a paper has more than ten authors). *Prince, A. F. K. et al.* does the job, but is a little unfair on the other authors (one of whom might be your examiner). It is best to include authors' initials, particularly with common names, such as Fisher, Wong or Patel, to make it clear exactly which Fisher, Wong or Patel the author is.

If the article has been published by a team or institution rather than individual authors, list the group by name:

West Cheam Landfish Taskforce (2015). Landfish Survey: West Cheam and adjacent areas. *Landf. Newslett.* **53**: 43–58.

Year of publication

This is essential. Remember, it is best to use the most up-to-date references, because this shows the examiner that you are following developments in your field.

Title of article or paper

This tells the reader about the contents of the paper – it can also be a useful last-minute reminder before an oral examination. Give all titles in their original language, with a translation in brackets if necessary.

Journal name

The journal name is sometimes the only way people can track down the reference, so you must include it. People usually do not bother to write the journal name in full because every journal name has a standard abbreviation, which you should use. If you are not sure of a journal name's abbreviation you can usually find out by checking the website of the journal in question.

Volume number, first and last page number of the article

The volume and page numbers help a reader to find the article. You do not normally need to include the issue number.

In order to find the article quickly the reader needs the first page number. Giving the last page number will make the examiner more inclined to believe that you at least looked at the article and did not just copy the reference citation from another paper. Check that your last page number is always greater than your first page number ...

For e-journals that are only available online, include all the information needed to find the article. Usually e-volume and page or article numbers are given on the first page of the article.

References from books

Follow the same principles when writing the reference details for books. When you reference books, cite the authors, the date, the title, the town or city where it was published, and the name of the publisher. For example:

Simon, C. and Mitchell, J. (2015) *A logic-based calculus of everyday objects.* (Hendrix and Joplin Press, San Francisco)

If you are citing a chapter from a book then include the above details, plus the name of the chapter, and the chapter authors, as well as the book title. For example:

Murray, K., Cain, C., Burrows, A. (2014) Successful research methods. In *Applications of Acoustics.* (Chard Press, Axminster)

If the book is edited then include the names of the editors in your reference details:

Weir, C., Gallagher, R., Snaith, M., Etienne, V., Robinson, C. (2016) Statistics don't mean nothin'. In *A Handbook of Paranormal Mathematics,* Stubbs, L., Fakir, A.D., Benson, R.O., Payton, L.R. (Eds). (West Cheam University Press, West Cheam)

You can also include the page numbers if you want to refer to just one section; *p* is used to refer to one page and *pp* to refer to a number of pages.

Gaye, M., Ross, D. (2017) Probability theory and gene cloning. In *Favourite Oncogene Stories.* (Robinson Publications, Detroit), pp. 103–128

Hewson, P., Evans, D., Clayton, A., Mullen, L. (2016) The growth of red giants. In *Stellar Evolutionary Theories.* (Lennox Press, Aberdeen), p. 206

Also list the volume number if there is one, and edition number if it is necessary for the reader to find the reference:

Karloff, B., Flowers, B. (Eds). 1959. Revitalisation of the Dead. Vol. 2. (Head and Throttle, Whitby)

If the publisher lists several cities – for example, *London, New York, Toronto, Sidney, Singapore, Cape Town, Delhi, Cairo, Brasilia* – you only need to list the first one.

Theses and dissertations

These are referenced in much the same way as books:

> Morissette, A. (2016) The effect of sharp-edged prophylactics on patient recovery. (PhD Thesis). University of West Cheam. p. 963

Conference abstracts and proceedings

Published abstracts and proceedings from conferences are referenced similarly, including the name of the conference and where it took place:

> Knight, G., Guest, W., Patten, E., Knight, M.B. (2013) Diffusion artefacts of scanning tunnelling electron microscopy. Fifth International Workshop of Electron Microscopic Techniques, Oglethorpe, USA.

> Harris, S., Murray D., Smith, A., Dickinson, B., McBrain, N, Gers, J. (2015) Disposal of the products of fermentation. Second European Conference on Catabolytes and Consciousness. Leyton, UK.

If the conference papers are published in a book or as a supplement to a journal, make sure these details are given, including page numbers. Be careful as some conference proceedings state that you are not allowed to quote material from the abstracts without permission of the authors.

Unpublished papers and talks

Give the speaker's name, date, title of the meeting, place of meeting:

> Morrison, M. Ballistics in a modern context. Paper read at conference on Psychology of Risk. (University of West Cheam, June, 2015).

If you are giving information that someone has told you and it is unpublished and cannot be referenced in any other way, then put 'personal communication', or 'pers. comm.' as a reference – but avoid doing this if possible as the reader cannot easily check your source.

Computer programs and the internet

If you have used computer programs, reference them fully: give the version number and reference the original paper in which the program appeared (or the name and address of the supplier of the program). If you have been searching databases then give either the version number of the database, or if the database is frequently updated give the date of your final search. If you have been using data or programs from websites give the name and address (http://www. ...) of each site.

The importance of referencing carefully

Referencing properly, although tedious, is not difficult. Never forget that your examiners are very, very human and that one of yours might have a particular bee in their bonnet about sloppy references (one of the authors of this book does). They could well have written their own thesis in the days before such fancy modern aids as reference programs existed, and might bear a grudge as they recall the hours they spent typing their thesis references when all their flatmates were out at a Sex Pistols concert. Just possibly.

Common Mistakes

- Citing a paper in the text, and forgetting to include it in the References or Bibliography section. Even with the use of reference database programs, losing references from the text is still a common sin. Some examiners will spot-check, so open a few pages and check that each reference that is cited is also present in the References or Bibliography section. You have been warned.
- Using incorrect journal abbreviations. There is no excuse for this because it is so easy to find the correct abbreviation. Examiners

can, and do, ask for corrections to errors in journal names, and this can be very time-consuming.

- Inconsistency in the style of citations and references. For example, these five references appear in the same References section of a thesis, each one comes from the same journal and each one is irritatingly different from the others:

Aguilera, C., Mayfield, C. The head-tilt feeding strategy of North American landfish (2016). Landfish Today 30: 132–139

B.J. Armstrong, M. Dirnt, T. Cool, & J. White. The Green landfish phenomenon. Landf Tod (2017) *32*, 67–73.

King BB and Wonder S. An integrated protocol for locomotor and musical testing of landfish. *Landf Tod.* 32: 8–12 (2017).

McCartney, P., Starr, R., Lennon, J., Harrison, G. (2013). The biology of northern landfish. *Landfish Tod. 23: 21–28.*

Springfield, M.I.C.B., *et al.* (2015). *Landf. Today* **28**, 21.

- Misspelling authors' names (especially on well-known papers). For obvious reasons, misspelling an author's name will count against you even more than normal if the author happens to be one of your examiners.

A friend of ours with name of Eastern European origin – although long not particularly difficult to spell – was asked to examine a PhD thesis. This examiner found his name cited throughout the text of the thesis but it was misspelt every time, in a variety of different ways. During the viva the student was made to suffer for his mistake and had to spend several days correcting all the offending citations.

- Getting the page numbers wrong. Guard against this by carrying out a quick scan of your references and check the last page number is higher than the first page number.
- Getting really famous references, whose details the examiners know, wrong; for example, citing the discovery of the structure of DNA in a paper by Wilson and Crick ...

Key Points

- Write your references into a reference database program from day one of your project
- Do not be lazy with references – statements made out of the blue, without the support of a reference, are obvious to experienced examiners
- Take care with references. It is very easy to make mistakes when typing them and examiners know this
- Use a consistent style for citations and references
- For journal articles enter authors, year, title, journal name with correct abbreviation, volume number, page numbers, according to the conventions of your field
- For books enter all authors, year, title, editors (if there are any), publisher and the town or city in which they are based
- Make sure your references are up to date.

Chapter 4

Planning and Writing Your Literature Review

4.1 Read Around Your Subject from Day One

When you start your PhD, master's or undergraduate project, there is an enormous amount of material that you need to learn so that you begin to understand what is going on. Most of all, you need to find out what your research group (however large or small it is) is doing and why. For this reason, your supervisor will give you articles and books to read and websites to look at. You will use these as starting points for finding other relevant literature to fill in the details about your field. At first it will be probably all be very confusing but eventually it will start to make some sense.

You will also have to find out exactly what goes on in the lab: what equipment is used; what experimental techniques are employed; how the data are assessed and analysed. If you are not carrying out experimental work (for example purely theoretical analysis, computer modelling or simulations) then you will still need to understand the techniques used.

At this point it is quite easy to get stuck into your practical work, perhaps at first helping a postdoc and then developing your own experiments, forgetting to keep on with your reading. This is a mistake because it is important that you continue reading around your subject, expanding your understanding of the field, and familiarising yourself with what progress is being made elsewhere. If you don't do this it will be hard to understand fully the motivation for the particular research that you have chosen to undertake.

Your supervisor should remind you to keep up with your reading but, regardless, it is *your* responsibility to make sure your background knowledge of the field is complete and up to date. When you start writing your thesis you need to have a well-developed understanding of your field.

4.2 Keep Notes from Your Reading

As you read different articles and books, it's important that you keep a record of what you have read and what you have learnt from it. If you do this efficiently it will save you from having to read it all again when you come to writing your thesis or dissertation. You will probably also want to keep copies of the most important articles.

People keep notes in many different forms. There are no rules about this except to make sure that you use a format that you are comfortable with and that allows you to keep a record of everything you need, efficiently.

Many people use a notebook for this purpose. You can jot down important points as you read them. Don't forget to include a record of where you found the information. When you start writing the thesis and refer to your notes, you will need to justify the statements you make in your thesis by citing the source of the information – for example a journal reference or a book.

Other people prefer to annotate copies of articles by highlighting important points or writing notes in the margin. This can be good because it's easy to see where your information came from and to link it to other relevant points in the article.

For articles from journals, the chances are that you have an electronic copy of the document. You might prefer to annotate the electronic copy rather than a paper copy because then it's easy to carry it around with you and you won't lose it (so long as you keep backups of all your files ...). Electronic copies of articles are good for lots of reasons, but one of the most important ones is that you can search them for keywords and this is much faster than flicking through paper copies of articles looking for particular words.

However you keep your notes, right from the beginning you need to set up a logical filing system for all the articles you use, so that you can find them again easily, whether on paper or in computer files.

It's also ESSENTIAL (we're putting this in capitals to attract your attention) that you take the time and make the effort to learn to use a reference database program/website and to import your references into your personal electronic library from DAY ONE (yes, really) of your project. Then you will have an electronic record of the bibliographic details for each reference you come across to enable you to identify it, whether or not you actually cite it. Most importantly, when you come to write your thesis or your own articles for publication, you will be very grateful for the time and effort you have saved by doing this as you go along rather than having to get all this information together at the end when you are in a rush to finish.

As we explained in *Chapter 3: Your References or Bibliography*, reference database programs exist to store information about references, automatically cite references and format references, and often you can write notes or comments for your own use in the personalised database you create.

4.3 Planning Your Literature Review

The first significant part of your thesis or dissertation is likely to be a Literature Review – perhaps in a chapter of its own or perhaps as a part of the Introduction. The Literature Review is there for a number of reasons. Firstly, it sets the scene and introduces the topic to the reader. Even if the reader is an expert in your broad area of science, they will probably not be so familiar with the particular aspects you have been working on, so you need to lead them into your work by showing them how it relates to the rest of the field. This is particularly important for the reader who happens to be your examiner, but it's also true for other people who read your thesis, for example a new student who is starting on their own literature search and who is wondering about the projects in your research group.

Secondly, you need the Literature Review so you can refer back to it when it comes to interpreting your own results and understanding how they how they relate to other work in the field. If you can't show its relevance to other researchers, then your work will appear to be on its own and its importance will be diminished.

Finally, for your examination it's very important that you can convince your examiner that you do actually understand the background to your own work and that you have not just done what you were told by your supervisor. A thesis or dissertation, at whatever level, must show that you know what your research is for and why it has been carried out. So a comprehensive and well-written Literature Review is a vital piece of evidence that you have a sound, up-to-date, understanding of why your research is interesting and important.

Since your Literature Review does not depend on your own results (they come later in the thesis), it's something that you can think about at an early stage. This means that you can take time to decide on a logical order for the material and to consider what to include and what not to include. You want to make it broad enough to set the scene for your work, but on the other hand you are not writing a textbook for your field, so you don't want to let it get too long and dwarf the original findings in your thesis. Similarly you want to go back a little way to show the historical development of the research, but you are not writing the complete history of science, so you need to be selective in including key discoveries.

As with all scientific writing, starting with a good plan is essential – even if it is just a list of bullet points. You should therefore take your time to develop a plan that gives a logical introduction to your field and then goes on to the specific areas that you are interested in, and the problems or questions that your research has addressed. This sets the scene for your next chapter, which will probably be about what types of investigations you have carried out, depending on the conventions of theses in your field and your department.

4.4 Writing Your Literature Review

If you have taken our advice, then when you finally come to writing the Literature Review, you will have all the information you need to hand ... You will have a set of references collected during the course of the project, with the citations safely tucked into your favourite electronic database program, waiting for you to use, and, most importantly, you will have a plan that is well worked out and includes everything you need to say, in a logical order. Make sure that you are also aware of the very latest developments by keeping up to date with the literature and finding out what is being presented at conferences.

You may find it difficult to get the right level when you write the Literature Review. You don't want to insult your examiner by starting at too low a level but at the same time a gentle introduction to the field – especially for a non-expert – is very helpful. We suggest you aim the start of the Literature Review at a scientist of your own level in the general area of your research. You will soon be able to move on to more technical details once you have set the scene.

Re-read your references if you have forgotten any important details. As you write, enter a citation into your text for all the significant statements you make. Don't be tempted to put in the references later – cite them as you go along and if you find that any are missing, don't leave them till later, just in case they say something different from what you remembered.

Once you have a draft of the Literature Review, get your supervisor to look over it if possible, so you can get feedback on your writing style. If there are any problems it is a good thing to sort them out early on. If you need to change the way you write, it's much better to find out before you have spent lots of time writing other parts of your thesis. If your supervisor doesn't have any major complaints, that will help you gain confidence about the rest of the writing process.

When you have completed your Literature Review, you will have made a significant step in producing your thesis or dissertation. You will have set the scene for what follows and with a bit of luck you will have whetted the appetite of your reader and left them wanting to read further to find out what you have added to this field of research.

Common Mistakes

- Neglecting to read around your subject while you are carrying out your research
- Not keeping a record of the papers and books that you have read
- Making a statement, and citing the wrong paper to support the statement
- Pitching your Literature Review at the wrong level
- Not using up-to-date references.

One of us was the PhD examiner for a student whose most recent reference was five years ago! This showed the student had made no effort at all to keep up their reading and they were given a considerable number of corrections requiring them to enter more appropriate references.

Key Points

- Start on your background reading early and keep up to date with progress in your field of research
- Keep comprehensive notes about what you have read
- Keep a comprehensive database of the references you have come across with all their bibliographic details in an electronic database program so you can cite straight into your text
- Write your Literature Review early on so that you gain some confidence in writing.

Chapter 5

Planning and Writing Your Materials and Methods/ Experimental Techniques

The aim of your Materials and Methods/Experimental Techniques chapter(s) is to produce a detailed recipe section for your experiments. By 'Materials' we mean the things you use for your experiments: reagents, theories and theoretical models, equations, simulation packages, plant or bacterial subjects, equipment, surveys, samples, existing data you are analysing, etc. By 'Methods' we mean the processes you use, whether experimental or theoretical.

Most disciplines, whether practical or theoretical, will have a section or chapter that fulfils this function, whatever the section is actually called: 'Materials and Methods', 'Experimental Techniques', 'Apparatus', 'Mathematical Concepts', 'Sampling Strategy and Methodology', 'General Procedures', 'Data Acquisition and Processing', etc. The conventions of this section vary widely between different sciences, for example, some chemists might not include a Materials section, whereas ecologists might separate this section into Processes rather than Materials and Methods. Whatever approach is normal in your field, the underlying principles are the same.

Although usually Materials and Methods will form a complete chapter, it sometimes makes more sense instead to have a chapter about each main experiment you carried out, which includes a section on the relevant materials and methods. This would apply, for instance, where different chapters address different questions with completely different approaches, giving unrelated results. In these cases you should be guided by what gives the most logical structure in your thesis.

Some theses, such as those written for certain medical degrees or theoretical subjects – for example, cosmology or mathematics – may in effect consist of a number of published papers sandwiched between an overall Introduction and Conclusion. In these cases, the Materials and Methods sections, whatever they might be called, will be included within each paper. Look at a good recent thesis from your department for your local customs.

Whatever the Materials and Methods/Experimental Techniques section is called in your discipline, and however it is arranged (whether as one chapter, or a section in each of the chapters or papers you are including in your thesis), your Materials and Methods have to be clearly and completely stated so that other people can use your exact methodology including precisely the same materials (whether they are chemicals, theoretical equations or other items) to replicate your results. You can assume some prior knowledge on behalf of the reader, for example, they will understand standard scientific terms within your field. If in doubt about how much prior knowledge to assume, err on the side of caution – ask your supervisor for advice and look at a good recent thesis or dissertation.

The importance of forward planning

Start writing your materials and methods in the word processing program you will use to write your thesis, as soon as possible – preferably from the first experiment in your project. If you do so you will have far less work to do at the end and will not be in danger of losing or forgetting important details.

If you have all your Materials and Methods in a computer file when you sit down to write the body of your text, celebrate! All you have to do is a little careful editing, and then you have finished this section!

If your materials and methods are only written in notebooks and various pieces of paper that have been jammed into files or stuck in a drawer or used to make paper darts for throwing at your friends, then start typing them out *right now*. Do not leave this job until last, when you will be most likely to make mistakes and might be starting to panic.

5.1 Planning Materials and Methods/Experimental Techniques

Include details of every experiment (or other pieces of work, such as simulations or calculations), for which you are giving results; be careful not to overlook minor experiments that are useful for supporting your main data. Include materials and methods for experiments that were unsuccessful or contradict your argument if they help you discuss your methodology and the main bulk of your results. Unsuccessful experiments will not count against you, as long as you can show that you have learnt from them – they may also be a useful warning to people not to spend time trying similar approaches.

Do not think that you must include materials and methods for every experiment you carried out during your project. If experiments are irrelevant to your final results they should not be kept in your thesis as they will only clutter your text and make it appear as if you do not understand the significance of your own data.

When you plan your Materials and Methods, group each type of material and each type of method and use sub-headings to separate your groups. Don't worry about the chronological order of your experiments, because this chapter is essentially the 'recipes' section for the rest of the thesis and you want all your 'cakes' to appear under one sub-heading, whereas your 'beers' will be under another and your 'pies' under the next, etc. Materials and methods should be presented in a logical order within the individual sections, for example, giving generally used materials, then specialist materials, followed by generally used methods, then specialised methods.

A common error: Confusing Materials and Methods with Results

One of the biggest difficulties for people who have to write Materials and Methods sections is confusing their contents with the Results sections. Materials and Methods are simply a set of instructions for the reader. Results are what you found out from your experiments, the data that you have generated. Sometimes it's helpful to have a look at some recent papers in your field, to understand how people separate Materials and Methods from Results.

Which to write about first, materials or methods?

Some disciplines favour presenting materials followed by methods, while others reverse this order; have a look at a good recent thesis in your field to find out the local conventions. Whichever order you present these sections in, it is generally easier to start by planning and writing your methods and to note down your materials as they crop up.

5.2 Writing the Methods Section

Write each method as if you were simply telling someone what you did. Be clear and do not get too wordy. We saw a thesis in which a student had written '... the weighing out of the agarose was undertaken until 5 g were measured out and the agarose was then later added to the solution'; the student could have more clearly and concisely written: '... 5 g agarose was added to the solution'.

You do not have to reinvent the wheel; but, particularly with more specialised techniques, go into as much detail as a reader needs to be able to repeat the process you used exactly. The small variations in methodologies between laboratories are often important. Make sure that you include any modifications to standard procedures that you have developed, particularly if you say your method used a kit/piece of equipment 'according to manufacturer's instructions' (see *Using kits*, below).

Do not explain too much about why you have used a certain method. You can put these explanations into your Results section.

Here are some, hypothetical, methods from different disciplines:

2.3.1 Preparation of 1-(tert-butyldimethylsilyloxy)-3-methyl-3-(phenyl-sulfonyl) propan-1-ol

Pyridine (5.2 ml, 55.7 mmol, 1.5 eq) was added to a solution of hydroxysulfone (11.3 g, 18.7 mmol, 2.5 eq) in CH_2C_{12} (70.7 ml) under neon at 4 °C. After stirring for 50 mins, TSB (12.7 ml, 27.8 mmol, 2.5 eq) was added dropwise. After 30 mins $CuSO_2$ (aq) was added to quench the reaction.

2.3.1 Wiksten–Bergmark–Schonfeldt–Karlsson–Malmquist condition

Let $f(x, y)$ be defined and continuous for all (x, y) in the region E defined by

$$a < x < b, \qquad -\infty < y_t < \infty, \qquad t = 1, 2, ..., n$$

where a and b are finite, $y = [y_1, y_2, ..., y_n]^T$ and let there exist a constant V such that

$$|f(x, y)f(x, y^*)| < V|y - y^*|$$

holds for every $(x, y), (x, y^*)$ in E. Then for every y_0 in $\mathbb{R}^n$ there exists a unique solution $y(x)$ of the problem

$$y(x) = f(x, y(x)), \qquad y(a) = y_0, \qquad a < x < b, \qquad f: \mathbb{R} \times \mathbb{R} \to \mathbb{R}$$

where $y(x)$ is continuous and differentiable.

2.3.1 Preparation of site for seeding

The field site was located at the half-acre field (University of West Cheam, Nonsuch, Surrey). An area 40.3 m × 15.0 ml previously sown to grass was ploughed and fertilised (Flowers–Icehouse Fertiliser) in mid-August 2015. The site was sown on 23 April 2015 with a spring crop (*Campanula cochlearifolia*) and netted to prevent theft or depredation by landfish.

2.3.1 Tallis–Byrd extrapolation

By this method the principal error functions of a given (low-order) numerical method are approximated, thereby accelerating its convergence. Suitable linear combinations of approximate solutions obtained by using the low-order method on different (uniform) meshes are calculated to complete this process. The Tallis–Byrd extrapolation depends on the knowledge of the powers of j appearing in the error expression for the low-order method.

2.3.1 Incubation of adults

Young adults collected from stock sub-cultures at 24.00 each day, according to the Monk–Williams–Gillespie–Pepper method, were placed in specimen tubes containing cotton wool at the base. Forty healthy adults were selected, covered with blue oyster extract (Cult Industries) and placed in each tube. The tube was then sealed with light aluminium foil. Tubes were placed in an incubator and kept at 37 °C and 85 % relative humidity for 24 hours.

2.3.1 Use of core samples

Conventional core samples from well bore 2 (Salisbury Avenue site, depth intervals: 233.87 to 243.23 m, 274.04 to 285.46 m) and well bore 3 (Cecil Road site, depth interval: 222.00 to 229.17 m) were taken. The samples were photographed and logged; records are given in Appendix 2. Samples were taken at intervals of 1 m.

2.3.1 The Roberts–Reid first order action

The normal Roberts–Dorge–Darvill action was taken as the model, but the metric and affine connections were considered as independent variables. The action:

$$S[g, \Gamma] = \int d^4x \, | \, g \, |^{1/2} g^{\mu\nu} R_{\mu\nu}[\Gamma]$$

was taken where Γ_{cd} is the affine connection. Then by varying Γ it follows that $\Gamma_{\mu\nu}$ is the metric affine connection defined by $g_{\mu\nu}$ and by varying g we obtain the Roberts–Reid field equations.

2.3.1 Transformation of bacteria

1 µl plasmid (1 ng/µl) was mixed gently with competent cells and incubated at 0 °C for 30 min. The bacteria were then diluted with 2 ml SC medium at room temperature and incubated at 37 °C for 45 min, shaking at 250 rpm. After incubation 100 µl of each transformation was plated onto LB agar plates containing 100 µg/ml ampicillin. Alternatively, *E.coli* Electro-shocker Transformation apparatus (Zappa Inc.) was used for electro-transformation of cells according to manufacturer's instructions.

2.3.1 Analysis procedure

A model equation, which best fitted the empirical data, was chosen and then a procedure to determine the coefficients of the equation was followed. Regression analysis was chosen in this study; because of the likely log-normal distribution for x (the weak interaction parameter) the regression analysis was carried out in the natural logarithm of x.

2.3.1 Preparation of a single ion

For experiments with single ions, it was necessary to eject unwanted ions from the trap. Following a loading sequence at a standard trap potential (50–100 V), the endcap voltage was pulsed to zero for a short time (10–20 µs) as soon as ion fluorescence was observed (typically after a delay of 5–20 s). If subsequently fluorescence from more than one ion was observed, the process was repeated until only one ion remained.

5.3 Writing the Materials Section

The Materials section is simply a list of the reagents or equipment you have used in your experiments (see Figure 5.1). This section may have different titles depending on your discipline. Most supervisors prefer you to start with commonly used materials and then list more unusual materials or equipment used for only particular types of experiments. Alternatively, you can simply list all your materials alphabetically.

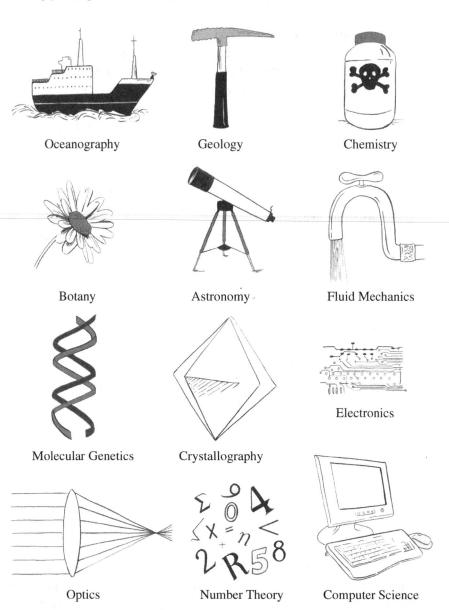

Figure 5.1　Some materials/apparatus.

As well as the materials and equipment, list your suppliers and any model or version numbers. For some items the products from different suppliers are more, or less, effective; therefore, someone attempting to repeat your experiments needs to know exactly what you used. When listing your supplier, you could include their head office city and country (or their website address) otherwise someone, particularly if they are working in another country, might not know how to contact them. If the address has been cited once, you do not need to refer to it again.

As with everything else, different fields have slightly different conventions. For example, a biology thesis may list the suppliers of even the most routine chemicals, whereas a chemistry thesis probably would not. The biologist uses the chemicals as a means to an end and relies on the supplier for quality control, but the chemist may well synthesise, purify and analyse their own chemicals. Ask your supervisor and check a good recent thesis in your subject if you are not sure what to list.

In the example below we show a typical Materials section from a molecular genetics thesis (alternatively, much of this information could be given in a table):

Chapter 2 Materials and Methods

2.1 Materials

2.1.1 General reagents
All laboratory chemicals were Analar grade from Springsteen Chemicals (Long Branch, USA) with the following exceptions: Trisma base, ethidium bromide and dextran sulphate were from Cobain Inc. (Seattle, USA); IPTG and X-Gal were from Garcia Industries (San Francisco, USA); sodium chloride, trisodium citrate and potassium acetate were from Cocker Plc (Sheffield, UK).

2.1.2 Bacteriological reagents and bacterial strains
Agar and agarose were from Garcia Industries. Tryptone, yeast extract and NZY broth were obtained from Minogue Pty. Ltd. (Melbourne, Australia). Antibiotics ampicillin (sodium salt) and tetracycline were from Fuemana Ltd (Otara, New Zealand). Kanamycin was from Cocker Plc and ultravox was from Ure (Cambuslang, UK).

E. coli strain DStH came from St Hubbins Plc (Squatney, UK), and the recombination deficient strains NT and DS came from Tufnel Industries (Squatney, UK) and Smalls Plc (Nilford-on-Null, UK) respectively.

In some research projects you may have made up solutions or other reagents (either standard or specialised) for which you need to give a 'recipe', in which case you can include a 'Reagents' or 'Solutions' section. You can write recipes for solutions in two different ways:

(1) State the volume and concentration of each reagent that is to make up the solution, and the final volume of the solution:

Bach solution:
200 ml 5 M sodium chloride
100 ml 1 M sodium citrate
water to a final volume of 1 l

If you write recipes in this way, remember that if you give volumes, you must also give concentrations; for example, writing 'add 200 ml NaCl' does not tell us how much NaCl to add, unless you have previously stated that the NaCl is at a concentration of 5 M.

(2) State the final concentration of each reagent in the solution:

Bach solution:
1 M sodium chloride
0.1 M sodium citrate

The five recipes below are written according to method (2), and so, for example, the solution '1 × TBE' is made to a final concentration of 45 mM Tris-borate and 1 mM EDTA. Remember to include the pH of any reagent for which this is important:

2.2 Solutions and media
2.2.1 General solutions

1 × TAE:	10 mM Tris, pH 8.0, 1 mM EDTA, pH 8.0
1 × TBE:	45 mM Tris-borate, 1 mM EDTA
20 × SSC:	3 M NaCl, 0.3 M trisodium citrate
Denaturing solution:	1.5 M NaCl, 0.5 mM NaOH
Neutralising solution:	1 M Tris, pH 7.4, 1.5 M NaCl

When giving the composition of reagents and referring to percentages, for example '20 % ether-petrol' or '0.6 % acrylamide', be clear whether you are referring to volume:volume or weight:volume percentages.

5.4 Writing Conventions

Each discipline has its own way of writing Materials and Methods, and your best sources of information are good recent theses or dissertations in your field, and published papers. However, there are certain general conventions to remember: scientific writing is denser than normal English and avoids many words we would usually use, particularly prepositions (*of, in, on*, etc.). So, for example, you would write '5 ml water was added' rather than '5 ml of water was added', or '10 nm diameter', rather than '10 nm in diameter'.

Take care not to use the colloquialisms of your laboratory or discipline; for example, a chemist would refer to methanol as CH_3OH, using standard chemical nomenclature that is recognisable by everyone, while a biologist might write the colloquialism MeOH which could be meaningless to other scientists. If you are going to use such colloquialisms, which are non-standard abbreviations, include them in your List of Abbreviations. Do not capitalise the names of chemicals; for example, Sodium Chloride and Sodium chloride are incorrect, sodium chloride is correct.

Remember to use the correct SI units and abbreviations for times, masses, lengths, temperatures etc. (see *Appendix 7: SI Units*). Similarly, make sure that you use the correct chemical symbols and formulae. Take care to put all variables in italics (e.g., $x, t, f(y)$) but to leave units (e.g., m, K, s, kg) and chemical symbols (e.g., He, NaCl) in normal upright font. Remember to use the correct capital and small letters. There should be a non-breaking space between a quantity and its unit (e.g. 5 g, not 5g and 15 °C, not 15°C) and also before a percent sign (e.g. 7 %, not 7%) – see *Chapter 13: Numbers, Errors and Statistics*.

If you feel the need to explain the principles underlying your use of certain materials (or methods) make sure you get it right. In a recent microbiology-related thesis a student wrote 'SM buffer is an enriched buffer which allows the growth of phage particles'. It does not: SM buffer allows the stable elution and dispersal of phage particles, which is something entirely different (this buffer does not support their growth). This kind of error is a clue to examiners that you are not really familiar with your field and may not fully understand your experimental techniques.

Which tense to use

You can write your Materials and Methods in either the present or past tense. The present tense can be used for general procedures that are exactly repeatable, the past for specific procedures that are not. In some cases, such as mathematical modelling, for example '**2.3.1 Tallis–Byrd extrapolation**' (p. 76) we gave above, the past or present tense is appropriate. In other cases, for example, the geology method, '**2.3.1 Use of core samples**' (p. 76) the writer would need to use the past tense. Once these core samples have been taken, exactly the same samples cannot be taken again and therefore writing in the present tense is inappropriate. We recommend that you read the section on scientific writing style in *Chapter 18: The Use of English in Scientific Writing*, Section 18.1.

Including the details

Include all the details of your protocols so another scientist could repeat them exactly. For example, if a solution needs to be of a particular alkalinity or acidity, or if the purity of a chemical is important (some of them are sold at different grades of purity) give the reader the details. You may also want to add useful information about how to make up the reagent/solution/medium, reminding people to filter, sterilise or autoclave solutions, for example:

2.2.2 Tchaikovsky agar
Add 20 g agar to 200 ml freshly squeezed sugar plum juice. Add 800 ml water and autoclave immediately.

If you are, for example, describing geographical areas, do so precisely, giving map references or GPS coordinates as appropriate, and using accepted spellings. If you are dealing with living organisms, give their full taxonomically correct species names, and give details of their sex, strain, age, diet and other environmental conditions, if this is relevant. Also, if appropriate, give the name of the authority or ethics committee that gave permission for the experiments; give similar information if your subjects were people.

For items such as optical, electronic or vacuum components that have been incorporated into a larger piece of equipment, specify exactly what was used by giving the specifications along with the name of the manufacturer and the part number.

All the quirky details of what you've done should be noted when writing up your Materials and Methods. At one time, a particularly well-known mouse genetics facility was only using a certain brand of pure spring water, available in supermarkets, to culture embryos. Other people had difficulty with their methods, until this important detail was revealed.

Calculations

If you include details of any calculations, use standard mathematical notation and put in enough explanation so that the reader can follow what you are doing easily. Begin each new stage of your calculation on a new line and use linking words like *and, so, therefore*, to guide the reader through your thinking. If you have a lot of data it is worth using a graph or table to provide a simple display for the reader.

5.5 What to Include

Specialised equipment

While it is not necessary to describe the use or setup of standard items of equipment, it is a good idea to include a section within your Methods covering any specialised equipment you have used (see Figure 5.2). Acknowledge workshops that helped you and give the specifications of apparatus you made or had made for you. Describe the reasons for any approach you took to building equipment or models: for example, economies of budget may have dictated the use of cardboard or

Figure 5.2   Specialised equipment.

plywood instead of acrylic or steel in the early 'cut and fit' stages. Also reference experimental designs that are based on previous work: if your vortex tube apparatus was inspired by that of Lake and Palmer (2015) and by the later models of Emerson (2016, 2017), then give them a reference and present your modifications to their designs.

Technical details – for example, calibration results for particular items of equipment – can either be included here or put in an appendix. What you choose to do with technical details depends to a large extent on the conventions of your discipline and how necessary these details are for the reader to understand your experimentation. If the calibration methods are non-standard or particularly error-prone you may wish to have a separate section for them as we discuss below in *Statistical analysis, artefacts and repeatability of measurements.*

Make full use of diagrams in this section. A picture paints a thousand words and it is often impossible for a reader to fully grasp a design from a written description alone (see *Chapter 14: Figures and Tables*). Block diagrams of an experimental setup, an electronic circuit or an experimental procedure can be particularly useful. However, detailed diagrams such as technical drawings or complete electronic circuits could go into an appendix.

Controls

Experimental controls are absolutely essential for checking the validity of data, therefore they must be included in your Methods.

Using kits

You do not need to copy the instruction manual for techniques in which you have used a kit providing you refer readers back to the manual. However, give a brief summary of how the kit works and refer to any papers containing the methods on which it is based, so the examiners can see that you understand what you have been doing – it is essential that you really do understand the principles that the kit is based on. For generally used protocols that occur within the kit write 'according to manufacturer's instructions'. You must describe in detail any additional steps or modifications that you have made to the kit. For example:

Full-length cDNA was prepared directly from tissue culture cells based on the method of Cherry and N'Dour (2014) with a SpeediNuck cDNA synthesis kit (Eagle Titiyo AB) according to manufacturer's instructions. The following modifications were made to the manufacturer's protocol: for the initiation of cDNA strand synthesis, the oligomer used had the sequence $(T)_{20}$; for the insertion of the double stranded cDNA into the vector, the ligation reaction was carried out with T53 ligase (Catatonia Ltd, Cardiff, UK) at a final concentration of 0.1 U/ml.

For unusual protocols or ones in which the manufacturer's manual has not supplied a reference to a published paper, you might want to put a detailed protocol in an appendix.

Statistical analysis, artefacts and repeatability of measurements

Errors exist in any experimental system. Always include a statistical analysis of your data and include estimates of errors, including any possible systematic errors. Discuss the reasons why you chose particular methods, and take care to carry out the appropriate calculations.

For some projects you may also need to include a section on artefacts. Artefacts are results that arise as a by-product of the methodology you have used, rather than being usable experimental data. A discussion of artefacts or other likely sources of error is particularly important for projects in which you have developed new techniques. Some experimental approaches may result in a skewing of the data, and you should think carefully if this applies to any of your work.

List any factors that might make it difficult for someone to repeat your experiment. This will be more of a problem with some experiments than others, for example, geological field studies may have many more problems of repeatability than a laboratory-based chemistry or physics project.

Computer programs and the internet

If you have written computer programs, these may fit best in an appendix. Reference any available computer programs that you have used, including their version number, by referring either to published papers or by giving the supplier. Give the details of any websites and databases that you have used.

Acknowledging other people if they helped you

If other people provided significant help for certain experiments, carried out part of an experiment, or provided samples that you worked on, you must acknowledge their assistance. Include your helper's name and their university or college if it is different from yours. You could write '... ethanol extraction experiments were performed with the aid of Mr R. Blaze' or '... samples were provided by Dr Harry

Judd and Professor Charles Simpson, Uppingham University', or '... resonance data for the Gabbitas analysis were kindly provided by Drs E. Minton and A. Davies', or '... wavelength calibrations were undertaken by Professor K. Melua of Nonsuch University'. Refer to *Chapter 11: Other People's Work*.

Figures, tables and appendices

By 'figures' we mean illustrations that are included in the text, for example line drawings or photographs of a piece of equipment, or flow charts of processes. It is often far easier to get the point across by using a figure rather than relying solely on a written explanation, so use figures wherever they will make things clearer (see *Chapter 14: Figures and Tables*). Take care when writing on the raw data from your experiments, such as X-ray films, traces, photographs, etc., which you might use later as figures in your thesis or dissertation. If you write notes on your data, do so neatly, so that you do not obscure any important details and so that the writing can be edited out if you need to use these data as a figure.

If you are using a particular notation to create your figures, such a chemists use to describe molecular configurations or electron orbitals, bear in mind that these are usually not standard, so explain your notation fully.

Tables are the most efficient way of presenting a large bulk of information, for example calibration readings from instruments, so make good use of them (placing them in an appendix if appropriate).

Some departments favour placing the materials and simple, generally used methods, such as computer programs, recipes for buffer solutions, proofs, etc., into an appendix. See a good recent thesis for guidance on your departmental conventions.

Reading what you have written

When you have written your Materials and Methods, try imagining you are a scientist in your field, but in another country, who knew nothing about the experiment, but had to repeat it. Could you do this from the instructions you have written? Get a friend who knows something about your field to read this section and see if they can follow it.

Common Mistakes

- Do not use different non-standard abbreviations in different places for the same materials or units, etc.
- Use the correct Greek letters such as α or β, rather than incorrect Latin letters such as a or b.

  One of us sat in on an exam in which an unfortunate student who had written ul for μl in their thesis was made by an examiner to replace every single ul in the thesis (several hundred in all) to μl. Beware!

- When writing chemical formulae (or other specialised nomenclature like nuclear isotopes or atomic energy levels) be careful always to use the appropriate subscripts and superscripts, for example:

$$C_{44}H_{62}O_6SSi_2 \text{ not C44H62O6SSi2.}$$

- When writing chemical names be careful to italicise appropriately, forexample 1-(*tert*-butyldimethylsilyloxy)-1-methyl-3-(phenylsulfonyl)-4-butanol.
- Check your English! For example, don't write '5 g gold dust were added', when the correct English is '5 g gold dust was added' (because you are referring to an *amount* of gold dust, which is singular). See *Chapter 18: The Use of English in Scientific Writing.*
- Do not use 'rpm' instead of '$\times g$' when describing centrifugation protocols. This is because every centrifuge is different and the force on the sample depends on revolutions per minute (rpm) and on the radius of the centrifuge. The expression 'times gravity' ($\times g$) specifies the centrifugal force, and is well understood.
- Do not use colloquialisms because your meaning may be mistaken, for example, never write 'hot' when you mean 'radioactive'.
- When you are describing actions, use the correct word. For example, do not use the word 'spin' if you mean 'centrifuge'.

We recently encountered a sentence that included 'when the sequence was fed into the database ...' Use correct English, and 'enter' sequences rather than 'feed' them. Another common mistake is to 'run' samples rather than (for example) 'electrophorese' or 'analyse' them. This may seem trivial, but again, one of us was examining a thesis and came across the phrase 'the tube was pulsed'. What does this mean? It transpired (in the examination) it meant that the tube was centrifuged at high speed for 2 s.

- Learn about the conventions of notation in your field. For example, there are globally agreed rules for how to write human and mouse gene names – human gene names are written in italicised capital letters (*DYNC1H1, SOD1*) whereas mouse gene names are written in italicised letters, only the first one of which is capitalised (*Dync1h1, Sod1*). Know the conventions in your field.
- When writing species names of organisms, convention dictates that these are written in italics, with a capitalised genus name and a species name that begins with a small letter; for example, *Homo sapiens, Mus musculus, Caenorhabditis elegans.* You can, after the first mention, abbreviate this to, for example, *H. sapiens, M. musculus, C. elegans.*
- When writing methods that use people's names, capitalise the first letter.
- Atomic mass can be stated in Daltons (or Da) and the unit name is capitalised because it is named after the chemist John Dalton. But when using SI units named after people (e.g. Newton) the unit abbreviation (N) is capitalised but not the unit name (newton).

Key Points

- Write a simple and technical account of what you did
- Do not include any results
- Always use SI units, with the correct abbreviations, without full stops – see the table of SI units in *Appendix 7*

- Non-standard abbreviations such as 'RT' (room temperature) must be listed in a List of Abbreviations at the front of your thesis or dissertation
- List exactly the correct materials and methods including, for example, any important details such as pH
- State exactly which piece of equipment you used, because this may also affect your results; give the name, model number, and manufacturer
- Give exact references to computer programs, databases and websites
- Make good use of figures and tables; they are the most efficient way to convey bulky detailed information.

Chapter 6

Planning and Writing Your Results

In your Results chapter you are giving a brief outline of each experimental strategy (not a detailed protocol), then telling the reader exactly what happened and what you learnt. You are not giving detailed procedures, because these are already in your Materials and Methods/Experimental Techniques chapter.

Results can take a number of different forms. Although we will mainly refer to experimental results in this chapter, the same principles also apply to any other types of results in your thesis, for example theoretical results, observational data, results from computer simulations or calculations, etc. If the focus of your project was building a major piece of equipment, your results may comprise tests or calibration of your equipment.

Your presentation of results will depend on the nature of your research. You might put your results in one or more separate chapters, or, alternatively, as a section within a chapter covering one aspect of your research. However they are presented and whatever they are called, your results are the core of your thesis. They show the examiner what you did and what you found out.

However, not all theses or dissertations, particularly in theoretical subjects, have a Results section. If new theories have been devised or existing theories or findings are being developed it can be more sensible to integrate them into the 'Discussion' chapter. Have a look at a good recent thesis or dissertation in your field and talk to your supervisor for further advice.

6.1 Planning Your Results Chapter

Remember that the whole point of taking time to create your plan is to get things right before you start writing the text in full. With a good plan in place, writing the text is much *faster* and *easier* than if you are not writing to a plan, or are writing to a vague half-formed plan. Trust us on this. We cannot overemphasise the importance of detailed planning of your Results section, because it's what the rest of the thesis or dissertation is built on. Planning what you're going to say in your Results is really really important! Really! It's worth the investment of time, even if you are late for a deadline and panicking about the amount of writing you have to do. Having a good plan also means you will avoid having to go back and re-edit work because you've changed your mind about the order or emphasis of the Results.

In the Results chapters, as in the rest of your thesis, you are presenting a story in the most straightforward and digestible way possible, so your examiners can easily see what you have found out. This means that your results need to be logically rather than chronologically arranged, so that both you and your reader can easily grasp their relationship to each other and their significance. For most people, unless you were incredibly meticulous in the initial planning of your project and everything went exactly according to plan, the order in which you did the experiments will not be the order in which you present your results.

You need to do two things to plan your Results: first, to clarify yours aims, and second, to arrange your results to support your aims.

Clarify your aims

To plan your Results chapter you need a clear idea of the aim (or aims) of your project. Your results need to be organised so they support this aim. Whether writing an undergraduate dissertation or a PhD thesis, your aim has probably changed slightly during the course of your research in response to either your own or other people's work. You might not have quite the results you were expecting, possibly someone beat you to it and you had to rethink your project in the light of their findings, or you could have stumbled across a much more interesting avenue of research along the way. Do not feel welded to your initial aims. Remember that your aims should match the presentation of your results as far as possible.

If your results do not seem to bear much relationship to your original aim, now is the time to re-evaluate your aims. We should distinguish here between the original aims you set out at the start of your project and the aims you will present in your thesis. What you write now are the aims of your thesis: what you set out to do in the investigations you report in your thesis. This may well be different from the aims of your research when you started.

Arrange your results to support your aims. To plan the order in which to present your results you need to know what results you have. Think about your project and what you have achieved. Read through your practical books and any other notes that are relevant to remind you of your experiments. Note the aim and result of each experiment in a file on your computer or on paper. Once you have made these notes, arrange your results into a logical order, grouping them according to the overarching aims of your project.

Worst-case scenario ... a bunch of meaningless results. Your results are what you have to work with, so the best thing to do is to accept the situation and then look at why your results appear to be meaningless. Possibly if you compare your research with established work in your field you will be able to explain why things went wrong. It may have been because of badly designed experiments or perhaps it was just bad luck – science is like that sometimes. If you provide a critique of your methods and carefully analyse and explain your results, perhaps giving ideas for an improved approach, you should be able to produce a reasonable dissertation or thesis. You may find you need a total overhaul of your ideas, in which case it is best to scrub your original aim from your mind and start again. It is a lot easier to do this with someone else, so go through your ideas with your supervisor.

6.2 Which Results to Include

Include all results that are relevant to your aim: the major results (around which you build each chapter or section), and the minor results, such as making or testing reagents, or calibration results, that support the major results. Remember to include all data from controls because these show the validity of your experiments.

Do not exclude results that contradict, or might contradict, your aim. (Honesty is not only good for the soul: it is also good for people's careers.) If you have contradictory results, attempt to explain them. The same is true for artefactual results produced as a by-product of the methodology. Contradictory and artefactual results can arise in all experimental systems and computational procedures. If your experiments did not work and you have managed to discover why, it is worth going into a detailed explanation, so that the examiners can follow your reasoning and other scientists can learn from your findings.

What not to include

Do not include results that have become irrelevant to your aims. If you include them you will not only waste a lot of ink and paper, you will also waste time – yours and your examiners', who will wonder if you have really thought coherently about the significance of your own research.

Just as you should never include results in your 'Materials and Methods', you should not include materials and methods in your 'Results'. Do not present any detailed protocols, just a brief overview of your approach to each experiment. You can refer back to the 'Materials and Methods' section for the detailed protocols (see *Chapter 5: Planning and Writing your Materials and Methods/Experimental Techniques*).

Including other people's data

Few PhD projects are the work of one individual alone and it is fine to include an account of the work of other people you have worked with, providing you clearly state that it came from a colleague – and, of course, you must have their permission to include their work.

Citing other people by name when you have used their data means you give credit where it is due and avoid any charges of plagiarism (and thus potentially failing your degree). Cite published or unpublished data, and if someone helped you with a technique or provided you with a reagent, a sample or a computer program, refer to them as well (see *Chapter 11: Other People's Work*).

Questionnaires and other relevant information

If you have used questionnaires to collect data, include a copy of the questionnaire.

6.3 The Order in Which to Present Your Results

To arrange your Results chapter(s) you need to decide on your main results, then decide on the order in which to present them. Start with the results that are the simplest and underpin your other work. Then, once you have set these down and are on solid ground, move to the next result, making sure it is supported by your previous work, and repeat the process. 'Walk' from result to result like this, setting out a logical pathway for your reader to follow (see Figure 6.1).

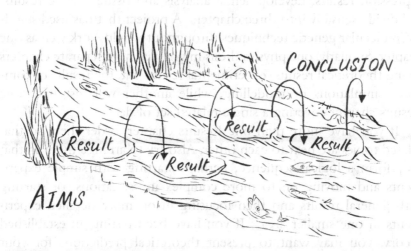

Figure 6.1   Walk from result to result, setting out a logical pathway for your reader to follow.

Arrange your supporting results into groups under your main results. Your supporting results are those from the many small experiments that you did in order to get the information you needed to do your main experiments. Present these results starting with the simplest followed by those that build on the data you have generated. Group each of these small results along with the main experiment they supported.

6.4 How Many Results Chapters, What Titles and in Which Order?

Once you have ordered your results into coherent groupings under each main result, then consider how many Results chapters to write. We have never come across any institution that has a rule about the number of Results chapters you are allowed, but do check your own local rules. Generally, the larger your thesis the more sensible it is to have a number of Results chapters simply because it is easier to digest the information if it comes in small well-defined chunks.

In a computing thesis, you might have two Results chapters, one describing tests of a program you have written, the second putting that program into action. In molecular biology, a project with protein expression results, developmental analysis and tissue culture results, will divide sensibly into three chapters. A project that has used standard molecular genetic techniques throughout would work well as one chapter. Similarly, one physics thesis might well have separate chapters giving theoretical results, experimental results and the results of computer simulations or modelling, while another will just have one Results chapter covering a single related set of data.

If you have more than one Results chapter, order them so that related topics are next to each other. Also try to arrange them so that they form a logical sequence, perhaps starting with simple experiments and moving on to more complex investigations, or starting with general results and then focusing in on more detailed experiments in one smaller area. If you have been testing an established theory, you may want to present theoretical predictions for your system first, followed by experimental results; on the other hand, if you have obtained new experimental results and then tried to develop a theoretical model to explain them it would make sense to put the experimental results before the theoretical ones.

Titles for Results chapters

You could simply call your chapter or section 'Chapter 3: Results'. However, a brief self-explanatory title would be much more useful to the examiner, especially if you have more than one Results chapter. You might want to write something like, 'Chapter 4 Results: Assessing the Pendragon method of processing liquid-sodium coolant' or 'Chapter 5 Results: Vaporisation of zinc alloys'. In fact, you do not have to use the word 'Results' in your title and could have a heading like 'Chapter 3: Mass spectroscopy of Palaeolithic bone samples', or, 'Chapter 5: Maser-derived observations of electro-magnetic emissions from Betelgeuse', or 'Chapter 6: 'Behavioural analysis of crimson landfish'. Exactly how you title your Results chapters will depend on personal taste and the conventions of your discipline.

6.5 Writing Your Results Chapter

When you have planned exactly how you want to present the results of your research, it will then be time to actually write your Results chapter. Your plan is there as a blueprint for your writing. You do not have to stick to it absolutely but you will waste a lot of time if you make many changes.

In your Results you are giving facts, not opinions. You are telling the reader which experiments you did and what happened. The prose of the Results chapter should flow smoothly from statement to statement so it builds rationally to support your aim. Make use of headings, figures and tables to break up the data into easily readable sections so that both you and the reader can keep track of your results.

When writing your Results state which method you used, and refer to the specific section of 'Materials and Methods' that describes this protocol. It will be helpful to the reader if you have arranged the materials and methods into the same order as the results – as much as is possible.

'Strategy' section and 'Summary' section

In some disciplines it is common to start the Results chapter(s) with a section explaining the experimental strategy used. This can be useful for setting the scene for your Results chapter(s) because it tells the reader the main results in advance, and makes it easier for them to grasp their significance. If you write a 'Strategy' section, keep it short and to the point. The reader does not want a detailed treatise on the philosophy behind the approach. Do not muddle the strategy with the results, which follow. A bullet point (or tabular) summary of findings at the end of each Results chapter is used in some disciplines and can be a useful way of reminding the reader about the main results.

Preparing for your Introduction and Discussion chapters

You will probably find that as you write your Results chapter you think of points that could go into your Introduction and Discussion chapters. Keep a note of these as you go along, either in a notebook or an 'ideas' file for future use.

Numbers and statistics

You may have to carry out detailed statistical or other numerical analysis of your data in the Results chapter. We cannot provide rules and guidelines for the hundreds of diverse types of number crunching that readers of this book will have to undertake. Your supervisor is the best person to comment on your individual needs. But there are a few general points to make about presenting numbers and calculations in any dissertation or thesis, which are discussed in *Chapter 13: Numbers, Errors and Statistics.*

6.6 Writing Style

It is probably best to write mainly in the past tense. If you use the present tense there is a danger of appearing to make unfounded generalisations, for example:

> ... a set of 36 related esters was chosen and analysed according to the Ofarim method: 6 of them were found to be positive, 28 were negative and no results were obtained for 2 of the samples.

This tells the reader your results – what you did and what happened. If we change this into the present tense:

> ... a set of 36 related esters is chosen and analysed according to the Ofarim method: 6 of them are found to be positive, 28 are negative and no results are obtained for 2 of the samples.

This sounds as if you are making a generalisation that for this particular experiment, 36 esters are always chosen and analysed by the Ofarim method, 6 are always positive, 28 are always negative and 2 never give results.

When presenting data, make full use of all the information that comes out of each result and explain its significance to the reader. For example, in a study of electrolytic corrosion, if you find that the average weight of two sets of alloy ingots has dropped, you could simply say,

> The mass dropped from 55.0 g to an average of 53.2 g for alloy A and 46.5 g for alloy B.

It is much easier, however, for the reader to follow if you also give an interpretation:

> The mass dropped from 55.0 g to an average of 53.2 g for alloy A, which was significantly higher than that of 46.5 g for alloy B.

Here we have not just presented the raw data to the reader: we have also worked with the data to make a useful comparison.

Use accepted scientific terms, and avoid jargon and scientific colloquialisms used around the laboratory. Use the correct units and their abbreviations. If you are unsure about the best way to present your results, try looking at published papers in your field for examples.

6.7 Figures, Tables and Appendices

Think carefully about which data to place in figures, tables and appendices because these can all be very helpful ways of displaying and summarising your results. Particularly in practical subjects, most of your results are in the form of data that could be presented as figures or in tables. Consider carefully what sort of presentation is most appropriate in each case (table, pie chart, histogram, graph, etc.). Your aim should always be to help the reader to understand the significance of the results and often it is really useful to show all your related results in a single table.

Your text is there to introduce and explain the figures and tables and to describe the data they contain. Do not reiterate the contents of the figure or table in your text: simply discuss the important points and refer the reader to the appropriate table or figure for more information. For a discussion of the use of figures and tables, see *Chapter 14: Figures and Tables* and for appendices see *Chapter 10: The Other Bits.*

Common Mistakes

- Including materials and methods in the Results chapter(s)
- Including duplicate or irrelevant results
- Writing results in a chronological not logical order
- Sticking with original aims, when the project has outgrown them
- Not making a plan first and rewriting the Results chapter(s) multiple times to try to get a logical order.

We have both, unfortunately, seen students waste a lot of time writing, rewriting and re-rewriting their Results chapters because they didn't make a good plan in the first place. Make a plan before you start writing!

Key Points

- Plan your Results by:

    clarifying your aims
    arranging your results to support your aims

- Arrange your Results logically and move from one result to the next related result
- Write multiple Results chapters arranged around different subject areas of your project if appropriate
- Make use of tables, figures and appendices to summarise your data
- Keep your style crisp and to the point; give facts not opinions
- A bullet point (or tabular) summary of each Results chapter can be helpful.

Chapter 7

Planning and Writing Your Discussion and Conclusion

In your Introduction you told the reader what you were going to do and explained the background to your research. In your Results you showed them what you had done and how you did it. Now in your Discussion you have to tell them what it all means: the relevance of your research and the conclusions that can be drawn from it. You should also outline future work that could be carried out.

In your Introduction you will start broadly by placing your work in a scientific context, then you will narrow down to state your specific aims. In your Discussion you start narrowly and broaden back out. Begin by restating your aims, then consider your methodology and results, and go on to build a wider picture of how your results fit into the context of your field of research. You could visualise your Discussion and Conclusion as the bell of a trumpet (see Figure 7.1).

7.1 Planning the Discussion

It really is a good idea to plan your Discussion thoroughly before writing it in full, otherwise you can tie yourself in knots. It is very painful to delete words and paragraphs you have laboured over deep into the night, however much they need to be discarded, and it is very easy to become mesmerised by your own words, especially if you are beginning to panic and are running short of time. When something

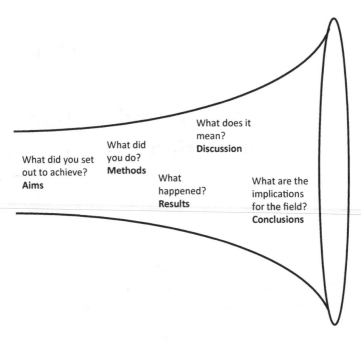

Figure 7.1   You could visualise your Discussion and Conclusion as the bell of a trumpet.

is written down in full, particularly if it is printed in a nice font and well laid out, there is a tendency to believe it more than if it were in note form. Examiners will not feel so indulgent.

While working on the plan of your Discussion chapter or section you might find you want to slightly rearrange the plan of your Results sections and Introduction. If so, do it. The whole point of planning is to get a blueprint that works. Be ready to go back and rejig your aims and Introduction plan if necessary, so that your conclusions grow rationally from your original question and the results of your experimentation or calculations.

Whether you have a single Discussion chapter, or Discussion sections at the end of each of your Results chapters, the Discussion section really should be a discussion of your project, and not simply a list of conclusions. Depending on the conventions for your field, your overall conclusions may be in a separate chapter.

Start by reading the plan of your Introduction and Results

Your Introduction and Results are the building blocks of your Discussion, so take time to read through your plan of them, and relate your work to the references in your Introduction and your Literature Review.

Start by considering your Results. Think about the details of your experimentation, why you chose particular methods, what each one told you, and whether the experiments or calculations could be improved. Then consider your results in relation to the wider field of your research – the information you have gained from your references and given in your Introduction. Take notes of all the points you want to include in the plan of your Discussion.

The beginning

Your opening paragraph should start by introducing your aims again:

> The aim of this project was to determine whether the increased incidence of dysfunctional adult behaviour among landfish in the Central Square area is caused by stray radiation from domestic television sets.

The middle

Now add points for discussion about individual results. Remind the reader why you carried out each experiment or calculation, and why you took certain approaches. Present minor conclusions from each experiment or calculation as you go along. You may also want to tell the reader how you could have improved the experimental aspects of your work, or your approach to theoretical problems. This is especially relevant if your project has not been particularly successful or a new technique has since been developed that would have been useful. Next relate your results to your field of research, citing the specific references which cover the area you are considering.

If possible your results should be discussed in the order in which they were presented in your Results chapter(s) – this will make your whole argument easier to follow. Make speculations if you wish, but only if they are soundly based in your research and are clearly distinguished as speculation rather than statements of fact.

Dealing with awkward points and comparisons

There may be, particularly with theoretical research, counter-arguments and alternative interpretations. Be aware of these and deal with them immediately they become relevant to your Discussion; alternative views will occur to your examiners immediately. If you leave these points to ferment in your examiners' minds they will be distracted from your argument, wondering if you have thought of the other possibilities. Acknowledge the problem:

> While there is some dispute about the existence of landfish in the A303 corridor (the Harrow Way) ...

and then deal with it straight away:

> ... I have found evidence of landfish spoors which point conclusively to their presence in the area around Basingstoke.

The end

Different disciplines and departments organise the end of the Discussion in different ways; your best guide is a good recent thesis or dissertation in your field from your department. Nevertheless, the same information goes at the end of almost all Discussions, however they are organised: there is a brief summary of the main outcomes of the research, a statement as to whether or not the original aim has been addressed, followed by suggestions for future work so the dissertation or thesis ends on a positive note.

In some disciplines this information is included at the end of the Discussion, possibly followed by a short final statement called Conclusion; in other disciplines it is put into a separate chapter or section called something like 'Concluding Remarks and Future Work'.

7.2 Writing Your Discussion

You will write your Materials and Methods, and probably also your Results, in a fairly terse and minimalist style. Your Discussion, like your Introduction, gives you a chance to write in a more descriptive and fluid style, explaining both what you have done and why you did it. Your prose should be easy to read, so lubricate it with linking words such as *and, therefore, but, however,* to join your ideas together. 'Signpost' your writing so the reader can easily see where it is going. Use headings for the main points of your presentation, and refer the reader to figures and tables in the Results, if that helps. Use the present tense for any general conclusions and the past tense to talk about your results. Try to write in the active because this is more direct and avoids wordiness:

> The laser radiation heated the sample.

> The ions passed through the cell wall.

rather than the passive:

> The sample was heated by the laser radiation.

> The cell wall was passed through by the ions.

Some people think that the 'scientific style' requires the passive; this is not so. Having said this, check to see what the conventions are in your field, as some examiners might feel more comfortable with a passive construction (see *Chapter 18: The Use of English in Scientific Writing*).

Figures, tables and appendices

Use figures and tables wherever they will help the reader understand your presentation. They are particularly helpful if you have a large number of comparisons to make, or wish to illustrate a point. It is unlikely you will have much need for appendices in your Discussion, but use them if you need to. See *Chapter 10: The Other Bits* and *Chapter 14: Figures and Tables*.

7.3 Writing the Conclusion

You will almost certainly have been working with an idea of the Conclusion throughout your project. As you planned your Results and Introduction it will have become clearer. Now that you have written your Discussion, your Conclusion should be crystal clear.

Make sure your Conclusion can be backed up by your data; it is best to do this before possibly wasting time writing a Discussion that does not in fact lead to your Conclusion. Plan, and look carefully at the results you have; they may be telling you things you have not thought about.

If you have answered the original question posed in your Introduction, then explicitly tell the reader so. If you were unable to answer the question then say why. If your work has specifically addressed any of the points from your references, make it clear what new light your Conclusion throws on these established findings or theories – note the main thrust of your argument at each point, making sure you include anything that either supports or contradicts your argument and making clear why it does so.

In some fields it is appropriate to have a separate short Conclusion chapter that pulls together the overall conclusions from the different aspects of your research and sets out how the work could be further developed. This is an opportunity to stand back and take a broader view of progress in your research field and where your own work fits in.

For some research projects, particularly those carried out over only a short period of time, you may appear to have no useful results. If this is the case for you, your Conclusion will be that you perhaps partially addressed your aim, or even that you could not address the aim of your project. This is perfectly valid providing you discuss intelligently why you were unable to address your aims. This situation usually occurs because of problems in the experimentation, either due to the experimental approach and design, or because of bad luck – the experiments just did not work in the time available, even though they should have done.

Suggestions for future work

Once you have settled on a Conclusion, jot down ideas for suggested future work and how you might approach it. Show that you can recognise the important questions arising from your work and have a realistic understanding of ways to address these questions. Don't forget to relate your work to other activity in the field.

Suggestions for future work are particularly helpful in upping your chances if you had a disappointing project. If you were unlucky with your results, or it was only towards the end of your project that you realised you were slightly off target with your research, or the project was ill-conceived and you did not realise this until it was too late to do anything about it, your suggestions for future work will show the examiners that at least you have learnt from your misfortunes or mistakes (which happen to everyone) and can see more productive avenues to follow.

Common Mistakes

- Insufficient discussion of the theoretical and experimental approaches taken
- Lack of organisation and insufficient flow from specific experiments or calculations through to the broader picture
- Not addressing key points raised in the Introduction
- Failure to present clear conclusions and relate them to your aims
- Not identifying future directions for the work.

Key Points

- Start your Discussion by outlining the general thrust of your argument: restate your aims
- Then review your results and their significance for the field
- Be aware of awkward points and deal with them immediately
- Finish with your Conclusion and suggestions for future work.

Chapter 8

Planning and Writing Your Introduction

Your thesis has to do more than simply supply the reader with data and a hint or two as to how you came to your conclusions. You have to take the reader firmly by the hand and guide them step by step through your experimental or theoretical research and the reasoning that brought you to your conclusions. The first step in that process is introducing them to your project, which is what your Introduction does.

In your Introduction you have to lay out the background to your research and show the relationship between your work and the wider field. You also have to tell the reader what question you have aimed to answer and why your project is interesting and important. You need to present your broad area of research and then narrow down to your specific interest, and finally pose the question that your thesis is answering. You could therefore visualise your Introduction as a funnel (see Figure 8.1).

8.1 Planning the Introduction

Do not hurry the planning of your Introduction. It sets the scene for the Results and Discussion chapters. A well-laid-out Introduction, along with well planned Results chapters, will make the Discussion a lot easier to plan.

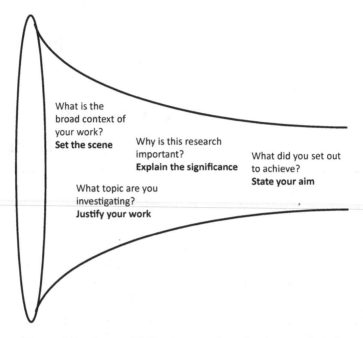

Figure 8.1   You could visualise your Introduction as a funnel.

Your Literature Review (or Literature Survey) may be included with the Introduction or it may be a separate chapter (see *Chapter 4: Planning and Writing your Literature Review*). Before you start planning the rest of your Introduction, make sure that you have completed the Literature Review. Look through any literature that might be relevant to your project so you do not miss any important papers. Track down any references you know you need, for example, the original reference for a method that you used, or the first paper announcing an important finding in your field.

When you have read through all your references and finished your Literature Review, you will have a thorough understanding of your field of research and this usually throws up new ideas about the significance of your Results. Note relevant points for your Introduction as you go through this process.

You need to have your aims clear in your mind before building an introduction to them. Note the key ideas you had while writing your

Results plan and the main points from your Literature Review. Use these notes to plan your Introduction. You will probably not get your plan right the first time around so your key points need to be in a form that is easy to arrange and rearrange. You could sketch your plan on paper, write each point on index cards and then shuffle them around until you get them in the right order, or shuffle the points around on your computer. At this stage, only pick out the major points.

Whether you are writing just one Introduction, or a series of smaller introductory sections to different Results chapters, the principles are the same: start broad, end narrow.

8.2 The Beginning

In your opening paragraphs you need to give the reader a precise but general overview of the area in which you are working. Cover any important findings or theories which led to your project or which affected your work, remembering to note any references you need. Bear in mind that a reader cannot ask you for explanations as they read through your text, so you need to be explicit and provide all the information you think a scientist in your field might need to understand your project.

There are two common ways of structuring the beginning of the Introduction. You could provide an account of your field at the time of writing (reviewing the current status of findings and theories, the *status quo*), or give a brief history of your field up to the time of writing (including the current status of findings and theories). As an example, think of a thesis written by a gardener who has been planting flowers as part of a landscaping project. The gardener has developed what they believe is the best combination of flowers and scents for a large garden. Using the first approach, their thesis would begin by saying what a garden is, then stating the architectural principles behind landscaping, and finally describing various types of flowers that could be planted: perennial, biennial, annual and so on. Using the second approach they would start by giving a historical introduction to gardens – the introduction of different types of flowers through the ages, and the first attempts at landscaping followed by the development of certain landscaping principles. Use whichever approach you prefer, *status quo* or 'historical', but the former – giving an outline of your specific area of study as it stands at the time of writing – is probably easier.

Remember you are introducing the reader to the small part of your discipline that you have been working in: you are not writing a textbook that is a History of All Science.

8.3 The Middle

In the middle section of your Introduction you have to take the reader from the general to the specific – your aims. You have to tell them what particular aspect of your field of research your project investigated and why. Give the reader an up-to-date outline of any findings and theories that are relevant to your project. Again, remember that a reader cannot ask you questions while they read, so you need to include all the information they will need to understand your project.

Strictly, every statement of fact you make should be referenced (see *Chapter 3: Your References or Bibliography*). In practice this is impossible, but reference all the key points and any ideas that are not either widely known or commonly accepted, so your examiners can check your sources. It is easy to under-reference, and difficult to over-reference a thesis. If you think you are in danger of either, get advice from your supervisor. You have to demonstrate constructive criticism of the references you are using, comparing and contrasting different theories and findings where appropriate. Although you should reference published papers wherever possible, you may also need to reference some websites or databases. If you do, make sure you give full details including the date accessed and or version number.

If it is possible, or relevant, the structure of your Introduction should reflect that of your Results chapter. Introduce the various aspects of your research in the order in which they are presented in your Results – using the plan of your Results as guide.

If your research concerns a new experimental method, piece of apparatus or computational procedure, then explain the problems with the old methods or apparatus. When you come to describe your improvements in your Results and Discussion, the reader will understand why an improvement was needed and why your approach is better.

Dealing with awkward points and comparisons

If you are working in an area where there are a number of competing theories, processes or findings, you will need to show the examiners that you are aware of all of them. If you put forward a theory which is perhaps quite controversial, show you are aware of the counter-arguments and alternative interpretations of data. When comparing complex theories, processes or findings it is best to go through the comparisons point by point. It is a lot easier to grasp the comparison if you run them side by side rather than setting out one theory in full followed by the other.

8.4 The End

The end of your Introduction is the simplest part to plan. Here you just need to tell the reader your aims – the question you are attempting to answer:

> The aim of this project was to determine if the increased incidence of dysfunctional adult behaviour among landfish in the Central Square area is caused by stray radiation from domestic television sets.

You could also give a brief outline of the structure of the rest of your thesis. A few simple sentences are all that is required. In particular, an introduction to your Results chapter may be useful: outline your experimental approach and the principal areas of investigation. If you have more than one Results chapter, note what is in each one so that the reader knows what to expect. You could put these notes in bullet points, which will look neat and be easy to understand at a glance.

8.5 Theory Section

Even if your project concerns an experimental topic, it may well be that you need to include some relevant theory. This is especially likely to be the case if you are working in the physical sciences. This section is concerned with how much discussion of theoretical aspects of your topic you need to include in your thesis or dissertation, and where you should put this theory.

The theory part of your thesis may be just a section of your Introduction chapter, or it may be a complete chapter that comes before your experimental chapters. Alternatively, if you have carried out a number of separate experiments concerned with different aspects of a topic, it may be more appropriate to have each of these experiments forming the subject of a single chapter, with the relevant theory included as a section within each of these chapters. In all these cases you should expect to come back to your theory at the end of the thesis in the Discussion chapter to review your results in the light of the theory you presented, and to draw any appropriate conclusions about the validity or otherwise of the theory.

It is usually the case in science that theory and experiment make progress side by side. There are times when theory is far ahead and experiment is trying to catch up, testing and evaluating theory that has been developed over a long time. On the other hand, there are times when experiment is far ahead and theory is trying to catch up, struggling to find a way of understanding a mass of related experimental results. But often they evolve together and benefit from each other's latest insights and developments. So the value of your work is enhanced if you too can help build links between theory and experiment in your own area.

How much theory to include

The amount of theory that you include in your thesis or dissertation depends very much on your topic and the kind of work that you have been doing.

If the focus of all your work has been experimental, and the relevant theory is well established and uncontroversial, you may only need to write a small section outlining the theoretical understanding of your area. The best place for it may be in your Introduction chapter or in the Literature Review.

On the other hand, you may need to carry out a significant amount of theoretical work yourself in order to apply the theory to your particular project, in which case you should present this work in your thesis. If the aim of your project is to test critically a piece of well-established theory, you will probably need to have an extended section of the Introduction or a separate Theory chapter where you outline the theory and its predictions for your specific experiments. If your experiments are intended to help resolve a conflict between two competing theoretical models of a subject, then you will need to spend time discussing the approaches taken in these different models.

Some people may have taken an interest in theory during their project and spent some time extending existing theories or even developing new ones. In cases like this it is entirely appropriate to have more than one chapter of the thesis devoted to theory.

Relate your theoretical section to your experiments

It is very important that the theoretical material you have included in your thesis or dissertation is relevant to the rest of your work. Sometimes students may feel that they have to present 'some theory' just because every thesis in their area always has a Theory section or chapter. But you shouldn't put anything into your thesis unless it has a purpose and forms part of the narrative of your research. So don't put in a Theory section just for the sake of it.

However, in the physical sciences it usually is appropriate to include a Theory section. The important thing is to make sure that it is relevant to the rest of your work. Don't include a discussion of a wide range of theory that's not particularly relevant to your research just because you know about it and want to impress the examiners.

But do make sure you have discussed and evaluated the theoretical understanding of your particular topic.

You may need to spend some time evaluating the validity of the theory for the experiment you are doing: are there approximations or assumptions that need to be made in order to apply the theory to your situation? Can you make any calculations or measurements that test the validity of the theory for your experiment? Are there other theoretical approaches that might also apply? These are some of the questions that examiners will be asking themselves when reading your thesis.

Finally, at the end of your thesis, after you have presented your results, relate these results back to the theory you presented at the start of your thesis. Does your work confirm or reinforce the theory? Does it raise questions about its validity? Does it allow you to decide between two different theories? Does it help point the way for future theoretical studies?

8.6 Writing Your Introduction

In your Introduction you not only have to make sure what you write is relevant, honest, rational and can be backed up by evidence, you also have to sustain the reader's interest through your argument. Your Introduction should give the reader a sense of setting off on a journey of discovery. You are taking them by the hand and leading them into the world of your research. They might have been there before, or they might not – even if you are taking them over fairly well-known territory, there is no reason the trip should be boring. You are interested in your subject and the background to it. There are lots of fascinating findings and theories, some of them perhaps quite controversial. To keep your reader's attention your prose needs a feeling of dynamism. Keep your sentences quite short; this will keep the energy up. Each sentence should follow logically from the previous one and towards the next. Similarly with your paragraphs: do not make them too long, and stick to one main idea in each one.

Remember that you are now an expert in your area of research. It is easy to forget that it has taken you a long time to get to this level

of knowledge of your field. Things that are obvious to you now probably weren't when you started, so be careful not to make too many
assumptions about what your reader already knows.

Try to avoid wordiness, and keep clear distinctions between statements of fact and statements of opinion. Include only relevant information. If you need to, define your key terms at the beginning of your
Introduction, and make sure you understand specialist terminology,
because its misuse will make the examiner question how well you
understand your subject and look out for more serious flaws in your
thesis or dissertation. For further advice see *Chapter 18: The Use of
English in Scientific Writing.*

Figures, tables and appendices

Use figures wherever they will clarify your argument. In the
Introduction you may wish to use figures and tables from published
papers or books; this is perfectly acceptable providing you fully reference them and providing you are not publishing your thesis and
so falling foul of copyright issues; see *Chapter 11: Other People's
Work*. Tables and appendices are the most efficient way of presenting
a bulk of detailed information. Make use of them. We discuss figures
and tables in *Chapter 14: Figures and Tables,* and appendices in
Chapter 10: The Other Bits.

Plagiarism

Never copy anything from other people's papers or work without
fully acknowledging and referencing it. Some writers think they can
get away with it here and there and cover themselves with other people's glory, but examiners know the literature as well as anyone and
are likely to recognise plagiarised work – it could even be theirs. It is
obvious when someone's writing style changes, which puts the examiner on the alert for plagiarism (see Figure 8.2). Students are rightly
failed for plagiarism. For more details see *Chapter 11: Other People's
Work.*

Figure 8.2   A student was found guilty of plagiarism ...

Common Mistakes

- Putting in too little information for the reader to understand the rest of the thesis or dissertation
- Not including enough references to support the statements that are being made
- Scattering information around the Introduction in no logical order, so that the ideas do not flow from paragraph to paragraph, or section to section
- Putting in technical terms without knowing what they mean (which can be embarrassing when the examiners ask you for an explanation).

Key Points

- Carry out a thorough and up-to-date review of the literature covering your area of research
- Ensure that you make proper reference to all ideas, results and figures that you use
- The beginning: start broad and set the scene
- The middle: narrow down to your chosen speciality, and set your work in context
- The end: ask one question that is addressed by your Results – this is your aim
- Keep your writing crisp, to the point and dynamic.

Chapter 9

Deciding on Your Title and Writing Your Abstract

9.1 The Title of Your Thesis or Dissertation

A short self-explanatory title is best. If it is too short the reader will not be able to tell what the thesis is about. If the title is too long the reader will have lost interest by the time they finish reading it. We will look at three possible titles for a thesis that addresses the question: *What are the courtship and reproduction rituals among northern landfish along the A645 road protection zone?* Here is the first possible title:

Landfish

Whilst this is a very compelling title, it is more suitable for a movie than a thesis (see Figure 9.1). It does not make clear what specific question the research project was asking, and implies that the thesis covers the complete field of landfish biology. This title is unhelpful to potential readers and in an oral examination the writer could justifiably be asked any question about landfish, rather than the one aspect of their behaviour that has been studied.

An Ethological Analysis of Sexual Behaviour, Courtship, Mating, and Reproductive Rituals and Rites among Male and Female Northern Landfish (*Gallus fritos*) of the A645 Road in

123

Figure 9.1 Choosing a good title can be difficult.

the Landfish Protected Zone between Kellington and Beal, Encompassing and Including the Embankments and Adjacent Regions including Hedgerows.

This is far too long and detailed. It is repetitive, unwieldy, and difficult to read. Cultivate a concise yet descriptive, efficient and accurate style in all your writing.

Reproduction rituals of northern landfish in a protected zone

This is just right. The title lets the prospective reader know what the thesis is about without going into redundant detail.

Whether or not you capitalise the main words of the title is up to you but generally it's best to use as few capitals as possible. If you capitalise all the letters, it looks as though you are shouting … so we think it's best to stick to the lowest possible number of capital letters.

REPRODUCTION RITUALS OF NORTHERN LANDFISH IN A PROTECTED ZONE

Oh, and one final thing: don't put a full stop/period at the end of a title: it would be redundant.

Thinking of a title

It can be difficult to come up with a good title. If you are lacking inspiration then try rewriting your aims. For example, if your research addressed the question *'How is the cognitive development of landfish between three and six months of age affected by external stimuli?'*, your title could be *'The effect of external stimuli on the cognitive development of landfish'*. If you need more examples, look at good recent theses in your field. Do not worry too much about sounding dry or stuffy as long as the title does its job. It is best to avoid redundant phrases such as *'A study of ...'*, or *'An investigation of ...'*. People should have realised it is a study or investigation of something by the fact it is being presented as a dissertation or thesis.

9.2 The Abstract of Your Thesis or Dissertation

Writing the Abstract is probably the first time you have had to summarise your entire project, but do not panic – you know your research better than anyone else.

In many ways your Abstract (or Summary) does the same two jobs as the trailer for a film or the blurb on the back of a novel – it says what your thesis is about in language that general readers can understand and it highlights the juicy bits that will make them want to read it. This is important for letting your examiners know what's coming, and also because many Abstracts are now available on the world wide web, you have the opportunity to tell other researchers across the planet all about your work.

The length of your Abstract

Sometimes you may be given a word or page limit for your Abstract, but often there are no rules as to its length and in this case it is a good idea to keep to just one side of paper. The Abstract is written for two audiences: for your examiners and for people researching in your area who may be interested in your results; they both need a concise and easily read *précis* of your work. By confining yourself to one side of paper you will necessarily keep to the main points, and satisfy both audiences. If the Abstract is too long, no one will read it ...

Planning your Abstract

Although your Abstract is only one side of paper long, plan carefully before writing it. Note the key points you wish to include. Start broadly by introducing your field of study, narrow down to describe your experiments, then assess the importance of your experiments and their wider significance. You could approach the task by taking notes in the following order:

- Field of Study: Introduce the broad area.
- Topic: Note your specific area of interest.
- Aim: What have you been trying to do or show and why?
- Experimental system or theoretical framework: How did you address your aim?
- Results: What have you achieved; did everything go as planned?
- Discussion: What new things do your results tell you?
- Conclusion: What significance does your work for your field of study?

Each of these bullet points could be turned into a couple of sentences of your Abstract. Keep your notes simple and only pick out the key details.

Writing your Abstract

Your Abstract will allow another scientist in your discipline who is not a specialist in your topic to get an idea of what you have done, so try to write in accessible language and avoid specialist terms as far as possible. The reader should be able to tell immediately from your Abstract if your thesis is of interest to them. Avoid using headings as they will clutter such a short piece of writing rather than make things clearer.

The contents of your Abstract should mirror those of the main text. As a very rough guide: your introductory section; your methods section; your results section; and your discussion section should all be roughly of equal length. Do not include figures, tables or references in your Abstract.

Writing your Abstract before you have written the thesis

For some degrees you may be required to submit the Abstract of the thesis well before you submit the thesis itself – PhD examining boards often require this so they can check what your subject area is. This same Abstract is often then placed at the front of the thesis, for your examiners.

How can you write an abstract of a document that doesn't yet exist? In this case you just have to do your best to anticipate what the content of your thesis will be. You should already have a good idea of what your main results and conclusions will be. It doesn't matter if in the end your thesis has a slightly different emphasis from that of your Abstract.

Common Mistakes

- Inaccurate titles

  Make sure your title really does describe the contents of your thesis. We came across one, otherwise excellent, PhD thesis suffering the indignity of having had its title slashed through with red pen. The author had been made to change the title even though there were only a few corrections in the main body of the thesis.

- Really long Abstracts

  Stick with one page. It's a good discipline to learn because scientific papers often have abstracts of no more than 200 words ... or fewer. Look at some papers to see how other people have written the Abstract.

Key Points

- Choose a title that is focused and best reflects your Results
- The Abstract is a short summary of your thesis – keep it to one side of a sheet of paper
- Don't rush the writing of your Abstract – this is the part of your thesis that will be read by more people than any other!

Chapter 10

The Other Bits

If you look at a recent thesis or dissertation in your field you will see that it does not simply start with page one of the Introduction. There will probably be an opening section made up of at least a Title Page and Table of Contents, which are there to help the reader navigate around the thesis or dissertation. You will need to include all or some of the following sections, and although there are usually no rules about the order in which they are presented, the layout below is often used:

Title page
Abstract
Acknowledgements/Dedication
Table of Contents (including Appendices)
List of Figures
List of Tables
List of Abbreviations
Main text of the thesis including References
Glossary
Appendices
Published Papers

You might use one or two, or all, of these sections in your thesis – it depends on the conventions of your field and the nature of your thesis or dissertation; we recommend you see a good recent thesis in your department for guidance.

Some people create a separate computer file for the sections at the front of the thesis, which precede the main text. If you do so you can edit it without altering the page numbers in the main text. You could distinguish this section by numbering the pages with Roman (i, ii, iii, . . . , etc.) rather than Arabic numerals (1, 2, 3, . . . , etc.). On the other hand, if you have your whole thesis as a single document your word processor can automatically enter the correct page numbers into the Table of Contents for you.

10.1 Title Page

We discussed how to decide on the title of your thesis or dissertation in *Chapter 9: Deciding on your Title and Writing your Abstract*. Each institution will have specific requirements for what else should go onto the Title Page of a thesis or dissertation. Check what these are in your department. Generally the Title Page requires your title, your name, your affiliation (department and institution), and a statement indicating for which degree you are entered, for example:

> Submitted in partial fulfilment of the requirements for the degree of Bachelor of Science of the University of West Cheam.

Some title pages will also include the month and year in which the thesis is submitted.

Figures 10.1 and 10.2 show typical Title Pages for an MSc thesis and a PhD thesis, respectively, that are being submitted to the University of West Cheam. Note where the text is, it's nicely spaced out so that you can easily see the title and the author and the other information that the University of West Cheam requires to be put on Title Pages. Note also that you do not put a page number on the Title Page.

**The application of mathematical models to
the problem of moving parts**

Stephen V.T. Tyler

**Submitted in partial fulfilment of the requirements for the degree of
Master of Science of the University of West Cheam
September 2015**

**Department of Mechanical Engineering,
University of West Cheam,
Nonsuch, UK**

Figure 10.1 Title Page from an MSc thesis that is being submitted to the University of West Cheam.

Identifying Genes and Related Pathways
Associated with Tricky Syndrome in Landfish

Adrian N.M. Thaws, MA

A thesis submitted for the degree of Doctor of Philosophy,
University of West Cheam
June 2016

Department of Neurogenetics,
University of West Cheam,
Nonsuch, UK

Figure 10.2 Title Page from a PhD thesis that is being submitted to the University of West Cheam.

10.2 Abstract

The Abstract (or Summary) is a short and succinct summary of what you have written in your thesis. It should explain what the thesis is about in language that general readers can understand and it should present the main results and conclusions of your work. For the benefit of people who find your thesis from keyword searches it should point out the new and interesting insights in your thesis that will make them want to read it. We discuss writing the Abstract in detail in *Chapter 9: Deciding on Your Title and Writing your Abstract.*

10.3 Acknowledgements/Dedication

This section is the most informal part of your thesis. It allows you to thank everyone who has helped you during your research project: your labmates and supervisor, parents or partner, friends who cooked you meals and bought you drinks. You may also wish to include people in your local engineering workshop, administrators or other people if they have been helpful. It is a good idea to be generous with your acknowledgements: this is the only bit of your manuscript that *everyone* is going to read and people remember whether or not they were mentioned.

Most people do not bother with a Dedication as well as an Acknowledgements section, but you might just wish to put in a short one-sentence dedication to someone or something special, such as your parents, partner or dog.

10.4 Table of Contents

This is exactly what it says it is. We have a Table of Contents at the beginning of this book – a list giving the page number of each chapter, heading and sub-heading. You can automate the creation of a Table of Contents by using an 'outliner' on your word processor which tags each sub-heading and automatically creates a Table of Contents (see *Chapter 2: Getting Organised*). If your thesis or dissertation is short, you can create the Table of Contents yourself if you prefer. The easiest way to do this is to (1) create a new file for the

Table of Contents, (2) make a duplicate copy of your finished thesis, (3) go through this copy and paste headings from it, into the Table of Contents file, making sure you note the page number for each entry as you go along. Never, never do this by working from the master version of your thesis: you are bound to have a word processing disaster if you do ...

You may have appendices at the end of your manuscript that present useful information that is too bulky or intrusive to include in the text, but which examiners need to know. The Table of Contents should include the appendices so that the reader can find this information easily.

10.5 Lists of Figures and Tables

The List of Figures helps the reader to find the figures they are looking for. Look through the final version of your thesis and list the figure number, the figure title and the page it is on. Graphs and charts are figures so include them in this table.

You should also have a separate List of Tables, by number, with titles and giving the pages they appear are on.

These lists can be generated automatically by some word processing packages, saving you a lot of extra work and probable mistakes.

10.6 List of Abbreviations

It is essential to provide the reader with a list of all the abbreviations you have used, especially all non-standard ones, so that there is absolutely no room for ambiguity. Many everyday abbreviations we use around the laboratory are non-standard and will not necessarily be understood by someone in another laboratory, let alone by an examiner from another university. For example, the abbreviation EtOH is well known to biologists, but may not be familiar to chemists or physicists or mathematicians.

Some people like also to incorporate the standard abbreviations, such as chemical or physical symbols, into their List of Abbreviations, but this is not necessary.

Keep a list of abbreviations as you write them in your text, and then put them into one list, in alphabetical order, with the long word or phrase that the abbreviation stands for. It's still best to define your abbreviations on first use in the text, even if you have included them in the List of Abbreviations.

10.7 Glossary

Most science theses won't include a Glossary, which is a type of mini-dictionary of specialist words. However, if the convention in your department is to include a Glossary then you could place it at the end of your thesis (for example, a thesis about a particular aspect of medical research might have a Glossary of specialist medical terms). If you are going to include a Glossary, add words to it as you go along. Then sort them into alphabetical order with explanations.

10.8 Appendices

Appendices are the sections at the back of your thesis where you can put large amounts of information that would detract from the flow of your main text, but which are still necessary to include. Here, put data and information that the examiner does not need in order to understand your argument, but does need in order to check your argument. Here are some examples of typical uses of an appendix:

- Tables of raw data
- Detailed results of database searches
- Calibration measurements
- Computer programs
- Technical drawings of apparatus
- User guides for software
- Lists of methods and suppliers
- Details of the error analysis.

You could also include published information, such as a well-known proof, that is necessary to support your work, to save the reader having to look it up in a library. Put different types of information into different appendices. Give each appendix a heading and number, and list them in the Table of Contents at the front of your thesis or dissertation.

If you are including a computer program in an appendix, print it using single line spacing and a small, but readable font, to save space.

Raw data or computer programs may take up so much space that you need to include them on a CD-ROM or DVD, in which case label these with your name, address, department and course, and ensure they are safely submitted along with the text of your thesis or dissertation.

10.9 Published Papers

If you have been an author on published papers that are relevant to your project, it is a good idea to include these papers with your text as they will help to show the examiners what you have achieved. Copies of the papers can be bound into the back of your dissertation or thesis. You may also include papers that are 'In press' (accepted for publication, but not yet published) or that are 'Submitted' (currently being reviewed for publication), providing that you clearly state 'In press' or 'Submitted' on the Title Page of the paper. Do not include papers that are 'In preparation'.

Common Mistakes

- Forgetting to put all the required information on the Title Page
- Inaccuracy in the Table of Contents – missing out sections of text or writing incorrect page numbers
- Forgetting to acknowledge someone who has given you significant help with your equipment or your state of mind while carrying out your research
- Not including appendices that would help readers to follow your work.

Key Points

- Check and fulfil any regulations concerning your Title Page
- Create a Table of Contents for the reader, and if appropriate include Lists of Figures and Tables
- Make sure you have a List of Abbreviations
- Put lengthy supporting materials into appendices at the end of your thesis.

Chapter 11

Other People's Work

In any scientific document it is normal to refer to other people's work. You may want to build on research carried out by other people, or compare your work to theirs, or even point out some errors. As we discussed in *Chapter 3: Your References or Bibliography*, you should always include a citation when you want to refer back to research by other people.

What we are concerned about in this chapter is using other people's work appropriately so that you act in a professional manner. This is a broad subject but we will discuss three main aspects: acknowledging contributions from other people in a team; plagiarism; and copyright.

11.1 Being Part of a Team

Most scientific research nowadays is carried out in teams. This may be a team consisting of a supervisor and a couple of students and/ or postdocs, or it may be a team of thousands of scientists such as the huge particle physics collaborations at CERN. This can lead to problems when writing up your thesis. If you worked closely with another student on aspects of your project, you need to make it clear what your contribution was, and what was that of the other student (and, of course, they will also have to do the same when they write their thesis).

Discuss with your supervisor how best to acknowledge the other people in your group. This may mean putting in a statement at the beginning of your thesis outlining how your work fits in with research carried out before you started or how the work you are reporting was shared between you and other team members. Or your supervisor may advise that the best thing to do is to indicate at the relevant point of your thesis where some of the work was done by someone else. Don't worry that this may weaken the thesis or make it look like you didn't do much. Your examiners will understand the situation – they probably have their own students working in teams as well. They just need to be able to focus on what part *you* had in the research.

In general, you won't describe work carried out by other members of the group in as much detail as the work that you carried out yourself. But you may still need to include some discussion – for instance, you may be in the situation where the apparatus you used for an experiment was made up of two parts, one of which you built while a fellow student built the other part. You will need to describe both parts but the discussion of the other student's part will be much briefer than yours, and you can include a note at the start of the section explaining that you didn't build this but the description is included for completeness. It would be the same if a project included two experimental strands, each of which was dependent on the other. Even if you were mainly involved in working on only one of these experimental strands, you would still need to include a short account of the other one. Similarly, if you simply inherited a complex piece of apparatus that was built by a previous student, you need to write your description of the apparatus in such a way as to make it clear to the reader what was already there and what you made yourself.

One of us prepared a set of questions about the apparatus described in detail in a PhD thesis we were examining. It was extremely frustrating to be told by the student 'I don't know about that because I didn't build it' when there was no indication in the thesis that this was the case. This did not get the viva off to a good start and of course it required changes to the thesis. Not acknowledging other people's contributions to your work is at best unprofessional and at worst it is dishonest and counts as academic misconduct.

11.2 Plagiarism

Plagiarism is using other people's ideas, text, data or figures without due acknowledgement. It therefore covers a very wide range of things from stating a fact without saying where it came from, to copying and pasting a large section of text from another thesis or paper without indicating the source. The first example may be defined as poor academic practice, with no attempt to deceive, and would need to be corrected before a thesis is finalised; the other is academic misconduct (cheating) and is completely unacceptable, possibly leading to failure of the degree.

Never copy anything from other people's papers or work without fully acknowledging and referencing it. Don't even think about trying this on: examiners know the literature well and are likely to recognise plagiarised work. It is obvious when someone's writing style changes, which puts the examiner on the alert for plagiarism. Students are rightly failed for plagiarism (see Figure 11.1).

It can sometimes be difficult to steer the right path between taking credit for new developments you have achieved and giving due acknowledgement to earlier work. Our advice in general is to err on the side of caution and always give credit to others unless you can clearly justify that what you have done is new and that you are solely responsible for it. But if you are in any doubt, discuss it with your supervisor.

It's worth noting that *self-plagiarism* is also bad academic practice. If you include in your thesis material that has been copied from one of your own papers, that is self-plagiarism and the source *must* be acknowledged. It does not make any difference that you were the author of the original paper. If you include in your PhD thesis material copied from a report you wrote as part of your BSc or MSc dissertation, it is still an offence because you are submitting the same material for credit twice. You may be able to use it if you state clearly where it has come from, but take advice from your supervisor.

One of us examined a thesis where large parts of one chapter were taken directly from a publication from the candidate's group. It interrupted the flow of the thesis because that material was not written for this

Figure 11.1   Think very carefully about acknowledging other people's work.

purpose, and the candidate had difficulty justifying what was written. This made for a very uncomfortable discussion in the viva and required major revisions to the thesis.

Your university is very likely to have documentation about plagiarism on its website. This will include advice about how to avoid plagiarism and how to acknowledge other people's work properly. It is well worth referring to this if you are in any doubt.

Some universities now make extensive use of anti-plagiarism software such as Turnitin. This will pick up overlap between your thesis or dissertation and any document on the web including published papers and websites. It's not worth the risk of being caught out by this software, which can be used on any type of report, including undergraduate project reports and PhD theses. A growing trend is for publishers to carry out these checks on articles submitted for publication.

11.3 Copyright

Sometimes you may not just want to refer to other people's work but you may want to reproduce some of it – perhaps a plot of earlier results in your Literature Review, or perhaps a lengthy quotation from a published paper. Generally this is fine, but you do have to follow the correct procedure, which means giving proper acknowledgement (of course!) and getting permission where necessary. This brings us on to the subject of copyright.

Copyright protects the intellectual property of an author in whatever form it is recorded: books, journals, photographs, compilations, offline databases, computer programs, websites, CD-ROMs, DVDs, etc. In the UK and the USA, for example, copyright lasts for 70 years after the death of the author. Copyright means that no one can reproduce an author's work without having first gained permission.

Although obtaining formal permission for reproduction of materials is always good practice, it is not essential for the printed copy of your thesis or dissertation that is submitted to examiners. This counts as 'fair dealing' ('fair use' in the USA), which allows for copyrighted material to be used without licensing under certain conditions. However, many universities now publish their theses online automatically in open repositories, and this means that they are much more widely available. For this published version it is essential to obtain the necessary permissions. Note that this may even apply if you are reproducing figures from your own publications, depending on the wording of the copyright form that was signed when it was published. Many publishers permit reproduction of figures or tables so long as you obtain the permission of the author, but you need to check carefully.

In the UK, for quotations, you can quote one extract of not more than 400 words from most publications, without infringing copyright. You are free to quote as much as you like from HMSO publications and the *Official Journal of the European Union*.

You can find further details about copyright on the websites of the UK Copyright Service (www.copyrightservice.co.uk) or the U.S. Copyright Office (www.copyright.gov).

A particular question may arise with copyright of text and images used in Wikipedia articles. You should check the site's documentation for details, but generally you can reuse textual material on a Creative Commons basis if you give a reference to the original source (but we *strongly* advise you not to copy text from Wikipedia). If you're unfamiliar with Creative Commons, have a look at this website: www.creativecommons.org. It's probably more relevant to consider using an image from Wikipedia. If you click on an image you will see the copyright information and that will tell you where it has come from and under what conditions you may reuse the image.

Check with your library if you are at all unsure about copyright restrictions.

Common Mistakes

- Failing to explain in what ways other people have contributed to your work
- Inadequate referencing
- Not obtaining permissions for reproduced material
- Poor understanding of what constitutes plagiarism.

Key Points

- Always make it clear what you did and what other people did when writing up your research
- Give a reference for every idea or fact that you take from another source
- Find out what permissions you require
- Don't copy and paste from other documents, even your own.

Chapter 12

Layout

A manuscript that is clearly laid out will be far easier to read than one thrown down onto the paper in any which way. Good presentation shows that you have taken care over every aspect of your project: if your manuscript is laid out logically and consistently it will help both you and your examiners to take your work seriously. It will also be easier for you to spot any omissions or mistakes when you are checking your drafts. When you are planning and writing your dissertation or thesis, the quality of your layout is nearly as important as the organisation of your ideas. If you make a good job of the layout, your thesis will look like a truly professional document, which is exactly what you want to convey to the reader.

Good layout needs good planning. Decide on the overall format of your document, the size of margins, headers and footers, line spacing, etc. before you start writing. Your department or university may have particular rules covering the layout of your text. Find out if there are any in your case, and if there are, act upon them – there may be a thesis template available. Changing the basic layout of your text halfway through or at the end of your writing can be very time-consuming. We strongly recommend having a look at previous dissertations or theses in your subject to see what you do and do not like about their presentation.

Aim to produce a document that is stylistically consistent throughout, looks well organised, and is pleasing to read. Here we consider basic formatting and layout.

At the outset, consider carefully what word processing program to use (see Section 2.6 of *Chapter 2: Getting Organised*). In particular, if your thesis is going to include a large number of equations (or other specialist display items and symbols) make sure that your word processor can deal with this sort of material comfortably. For heavily mathematical subjects, many people choose LaTeX, but this is not essential. Take advice but decide on what to use before you start writing.

12.1 Fonts and Line Spacing

Your choice of font and line spacing will affect the number of pages in your final manuscript. This does not matter as long as you have kept to your word limit if you have one (word limits are usually set for undergraduate and some MSc theses or dissertations, but not always for DPhil and PhD theses).

Fonts

'Font' basically just means typeface. Each font has a name to identify it, for example:

> This is Times font
> **This is Arial font**
> This is Calibri font
> This is Garamond font

Your word processing program will have a number of fonts available in different sizes. The size of your characters is usually measured in points. Very roughly, a single point is equivalent to 0.35 mm.

Choose a font that looks clear and is easy to read when it is printed (we do not recommend fonts like 𝔒𝔩𝔡 𝔈𝔫𝔤𝔩𝔦𝔰𝔥 𝔗𝔢𝔵𝔱; see Figure 12.1). Some fonts look good on the screen, but are not so pleasing on the printed page, and vice versa, for example, **Geneva** looks much better than **Helvetica** on some computer screens, but is not as readable when printed on some printers.

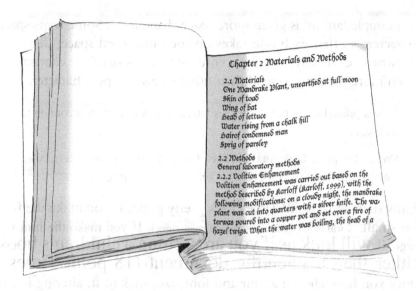

Figure 12.1 Use a font and layout that are clear and easy to read.

When choosing a font, type a phrase or two to see how it looks both on screen and on paper. Try 'The quick brown fox jumps over the lazy dog', which contains all the letters of the alphabet. Here are a few fonts in different point sizes:

The quick brown fox jumps over the lazy dog. (Times, 10 point)

The quick brown fox jumps over the lazy dog. (Times, 12 point)

The quick brown fox jumps over the lazy dog. (Calibri, 10 point)

The quick brown fox jumps over the lazy dog. (Calibri, 12 point)

```
The quick brown fox jumps over the lazy dog.
(Courier, 10 point)
```
```
The quick brown fox jumps over the lazy
dog. (Courier, 12 point)
```

The quick brown fox jumps over the lazy dog. (Arial, 10 point)

The quick brown fox jumps over the lazy dog. (Arial, 12 point)

The quick brown fox jumps over the lazy dog. (Garamond, 10 point)

The quick brown fox jumps over the lazy dog. (Garamond, 12 point)

Note how much difference a small change in font size makes! Most fonts space the characters according to their individual size, so, for example, an 'm' is given more space than an 'l'. Some fonts space characters so that each one takes up the same sized space, like on a typewriter. Generally, spacing according to the size of the character is much more pleasing to read than using one space per character.

This is printed using the font Times which spaces characters according to their size.

```
This is printed using the font Courier which gives
each character the same amount of space.
```

Fonts such as 12 point Times look pretty good. If you make the font too small, it will be very difficult to read (8 point Times). If you make the font too large, it will look as if you are writing a children's book rather than a scientific document (18 point Times). Once you have chosen a font and font size, stick to it: altering it will throw out the formatting of your text and you will have to spend ages reformatting.

Chapter and section headings will of course be larger than the rest of the text but figure and table titles and legends are often printed in a smaller font (and with wider margins) than the main text. Choose your font and make sure you are consistent!

Line spacing

Word processing programs use three standard options for line spacing. Single:

We amplified a 984-bp *Inxs* ORF from landfish head cDNA (Fig. 2.3). A second amplification product of 250 bp was also detected and this was sequenced, revealing a shorter splice variant.

One-and-a-half:

We amplified a 984-bp *Inxs* ORF from landfish head cDNA (Fig. 2.3).

A second amplification product of 250 bp was also detected and this

was sequenced, revealing a shorter splice variant.

And double:

We amplified a 984-bp *Inxs* ORF from landfish head cDNA (Fig. 2.3).

A second amplification product of 250 bp was also detected and this

was sequenced, revealing a shorter splice variant.

Your department or university will probably have a rule about the line spacing of your manuscript; usually double or one-and-a-half line spacing are specified as they are easier to read (these are also easier to edit when writing corrections on drafts).

12.2 Layout of Text on the Page

Margins, headers and footers

Your text should look comfortably settled on the page, so you need reasonably sized margins, headers (the space at the top of the page) and footers (the space at the bottom of the page). Increasingly, theses are printed on both sides of the paper – this saves paper. Most universities and colleges have rules about margin sizes and whether you can print your thesis double-sided or not. Find out if there are any in your case, and make sure you follow them. About 3 cm is a reasonable width all round except that you will need slightly more space on the inside margin so you will not lose any text in the binding: set a margin here of at least 4 cm.

You can put text in the header and footer. Some people type the chapter title or number in a small font in the header or footer zone so the reader can tell roughly where they are when flipping through their manuscript.

Inserting page numbers

You must number the pages of your thesis or dissertation. One of us, we will not say who, did not bother to find out how to automatically insert page numbers into their text when they first started using a word processor. Hours were spent writing them in by hand after the text had been printed. So find out how to insert page numbers. If you do not put page numbers in and the pages become mixed up after being dropped on the floor, it is a nightmare trying to get them back in the right order. You can put page numbers at either the top or the bottom of the page in your header or footer zone – there are no rules – but make sure the number is easy to see; avoid the left side of the page, close to the binding.

You can leave the first page blank if you prefer and number from page one of the Introduction, or from the opening sections. Do not number your Title Page.

Aligning to margins

There are different ways you can align, or 'justify', your text. With the majority of your text you will probably align it to the left margin:

2.3.1 Cook–Slater–Graham extrapolation

By this method the principal error functions of a given (low-order) numerical method are approximated, thereby accelerating its convergence. Suitable linear combinations of approximate solutions obtained by using the low-order method on different (uniform) meshes are calculated to complete this process. The Cook–Slater–Graham extrapolation depends on the knowledge of the powers of j appearing in the error expression for the low-order method.

or possibly so that all lines start at the left margin and finish at the right margin. This looks very neat and is generally the layout used in books and journals:

2.3.1 Cook–Slater–Graham extrapolation

By this method the principal error functions of a given (low-order) numerical method are approximated, thereby accelerating its convergence. Suitable linear combinations of approximate solutions obtained by using the low-order method on different (uniform) meshes are calculated to complete this process. The Cook–Slater–Graham extrapolation depends on the knowledge of the powers of j appearing in the error expression for the low-order method.

You might want to align a heading to the centre of your text. If you do this, only align the heading, and not the text underneath it as we have done below:

2.3.1 Cook–Slater–Graham extrapolation

By this method the principal error functions of a given (low-order) numerical method are approximated, thereby accelerating its convergence. Suitable linear combinations of approximate solutions obtained by using the low-order method on different (uniform) meshes are calculated to complete this process. The Cook–Slater–Graham extrapolation depends on the knowledge of the powers of j appearing in the error expression for the low-order method.

It is also possible to align to the right margin: this can be useful within tables but looks very odd in your main text (so don't do it):

2.3.1 Cook–Slater–Graham extrapolation

By this method the principal error functions of a given (low-order) numerical method are approximated, thereby accelerating its convergence. Suitable linear combinations of approximate solutions obtained by using the low-order method on different (uniform) meshes are calculated to complete this process. The Cook–Slater–Graham extrapolation depends on the knowledge of the powers of j appearing in the error expression for the low-order method.

Indenting

Indenting just means moving a section of your text in from the margin a little, like we have done for all the examples in this book. The easiest way to indent is to use the automatic tab settings on your word processor or to set a tab at the point to which you want to indent (see 'Tabs' in your online help manual for your word processing program).

When you are typing, all you have to do is hit the tab key and the cursor will move to the tab point. The tab only indents one line. If you need to indent several lines of text, move your margins at these sections rather than having to tab each line that you type.

For new paragraphs, you have to either put a line space to indicate a new paragraph is coming up, or you start the new paragraph by indenting it between five and ten character spaces. If not, use a tab key at the start of the paragraph to create the indent. The first line of this paragraph, for example, is indented in this way, and this is done throughout the book.

Alternatively, as mentioned, you need not indent, but can leave a line between paragraphs, as we have done here. However, this may make your text rather long. Whatever you do, you must clearly separate your paragraphs to make it easy for your reader to see the structure of your text.

Indenting quotes and examples

To distinguish quotes and examples from new paragraphs it is conventional to indent the left margin of the complete quote by about five character spaces, for example:

> This is the sort of indentation you might use when giving a quote. The whole paragraph or section has been indented by changing both the margins here.

Alternatively,

> For this paragraph, we have moved the margins in to indicate that we are giving a quote, and we have also used the tab key on the first line to indicate that it is a new paragraph.
>
> This is useful if you are writing a long quote made up of many paragraphs or sections, because you need to show that they are all separate.

Using different margins and indents

If you play around with your indent and margin commands (read the online help manual of your word processing program) you will find commands that allow you to indent the top line a different distance from the rest of the paragraph. This is particularly useful for formatting lists of references where you want to ensure each entry is clearly visible and separate, but you do not want to waste space by putting extra lines between the entries. For example, here the first line starts at the left margin, and all subsequent lines are indented five spaces:

Alexander, G. (2016). New radicals found in landfish. *J. Landf. Nutrition* 22: 59–67.

Harket, M., Furuholmen, M., Waaktaar-Savoy, P. (2017). Hunting high and low for landfish. *Landf. Tod.* 32: 119–125.

Illsley, J., Withers, P., Knopfler, D., Knopfler, M. (2015). Industrial diseases of landfish. *Landf. Disorders* 115: 15–31.

Isley, R., Isley, E., Isley, R., Isley, V., Isley, O., Isley, M., Jasper, C. (2017). A world harvest for landfish. *J. Landf. Nutrition* 24: 8–12.

Morrison, V. (2017) Brown eyed landfish. *Landf. Disorders* 122: 21–29.

Rahman, A.R. (2016) Landfish on the cricket pitch! *Landf. Newslett.* 54: 528–766.

Small, H., Pickering, M., Heard, P., Shovell, A. (2015). Don't look any further for landfish. In *A Handbook of Laboratory Techniques for the Study of Landfish*. (MUT Press, Cambridge).

Somerville, J., Bronski, S. (2016). Small town landfish. *J. Landfish* 108: 844–912.

12.3 Titles and Headings

Format

There are no rules about how you capitalise the words in your heading, but make sure your headings are consistent. Decide on a format and stick to it: use the same font and keep all headings in the same tense and otherwise stylistically similar. Most word processors will have a default style that automatically formats titles and headers appropriately. However, you can always change these if you want to. For headings and major titles you can capitalise the first letter of words except prepositions and conjunctions (unless they are the first word of the title):

The Principles of Para-Psychology in an Urban Setting

although many word processing programs have commands that capitalise the first letter of every word in a title or heading:

The Principles Of Para-Psychology In An Urban Setting

Your headings need to stand out from your text, but do not go over the top playing around with your fonts, otherwise your titles and headings will look messy and out of place with the rest of the text. The style we favour is for titles to be the same font, in bold, and either the same size or a couple of points larger than the text:

This is the Title, in Bold 14 Point Times
This is a Major Heading in Bold 12 Point Times
This is a minor heading in bold 12 point Times

Alternatively, you could use italics:

The Principles of Para-Psychology in an Urban Setting

Underlining looks fussy and we do not advise using it:

The Principles of Para-Psychology in an Urban Setting

We also do not recommend the use of capitals because they look as if you are shouting at the reader, and no one likes being shouted at:

THE PRINCIPLES OF PARA-PSYCHOLOGY IN AN URBAN SETTING

Shadowed or outlined looks like a party invitation rather than a heading:

The Principles of Para-Psychology in an Urban Setting

Having said all this, you can use any combination that you think looks good. We do not think this one does:

The Principles Of Para-Psychology In An Urban Setting

If a number forms part of your title it is best to write it out: *one* rather than 1, and *ninety-five* rather than 95.

An Experiment on Five Undergraduates Involving Electricity

Whichever style you use, bold, italics or the style of your choice, do not mix styles in different headings as this looks very messy. Note that it is not usual to put a full stop at the end of your heading or title.

Numbering: The title, headings and sub-headings

Do not number your title. In science theses or dissertation it is normal to number the chapter and section headings.

When you divide your chapters into sections, make use of sub-headings. To make it easier to follow the division of the headings and to help create the Table of Contents, sub-headings are usually numbered. The easiest and most logical way of numbering is by the chapter number, followed by the number of the sub-heading, for example, the first sub-heading of Chapter 1 would be 1.1, the next sub-heading would be 1.2, etc. Within these sub-headings you can further subdivide to 1.1.1 and 1.1.2, etc. as necessary. Do not put a full stop after the number. In the following example, the different categories of heading are indicated in italics:

The Development of the Cheam Moog Genesis Technique *(thesis title)*
1 Problems with traditional Moog Genesis *(first chapter heading)*
  1.1 The Cox paradox *(first division)*
  1.2 Instability in parallel Mackenzie fields *(second division)*
  1.3 Intermittent Cunnah emissions *(third division)*
    1.3.1 Roberts modulation *(first sub-division)*
    1.3.2 McCarthy wobble *(second sub-division)*

It is best not go beyond sub-sub-divisions (i.e. not beyond 1.1.1.1 or 5.4.6.3 to 1.1.1.1.1 or 5.4.6.3.7). You do not need to provide a numbered sub-heading for every tiny point you are making. If you really feel that you absolutely have to put in another sub-division and therefore another heading, then do not number the heading.

Separating sections and chapters

New chapters should start on a new page. Each new section should have a heading and be clearly distinct from the previous section, although it does not need to start on a new page. Never have a heading on the last line of a page.

12.4 Additional Features

Highlighting text

Occasionally you may need to highlight a word or phrase within your text. Using a **bold font** is usually the neatest way of doing this, but you could use *italics*. Underlining tends to look messy. Do not highlight words too often or they lose their impact and your text looks cluttered.

Bullet points

Most word processing programs allow you to make lists of bullet points (or numbered lists). Bullet points are used to distinguish different points you wish to make, for example, in a summary:

My research project entailed:

- using a Moog Genesis method to place Hsa21 into mouse ES cells
- injecting two of the resulting cell lines into host blastocysts to create chimeric animals
- breeding chimeras to create two independent lines of transchromosomic mice
- phenotypic characterisation at the behavioural and physiological levels of both transchromosomic mouse lines.

There are no rules as to whether bullet-pointed items should finish with full stops, but they often look messy if they do (we do not use them in this guide except after the final bullet point of a list). However, if each bullet point consists of several sentences rather than just a single phrase, it may be best to use full stops at the end of each one. Don't overdo the use of bullet points: for a formal document like a thesis they should be used sparingly.

Footnotes

Footnotes and endnotes are used to give additional information not strictly necessary to the argument. Footnotes are given at the end of the page,* and endnotes are given at the end of the chapter,[i] or in a list at the end of the book. Keep footnotes and endnotes to a minimum as they can be distracting to read. If the information is important, include it in the main text; if not, think about whether you need to include it in your dissertation or thesis at all.

If you are using numbers to mark citations, it is best to use symbols (*, †, ‡, §, #, etc.) rather than numbers to mark your footnotes and endnotes, otherwise they might become confused with your reference citations. Put the symbol in superscript (above the normal line) so that it is not read as part of your main text.

Check your layout on the final version of your manuscript

Print one final version of your thesis or dissertation, and skim through this checking that all your formatting is consistent and you have no errors, such as leaving a heading at the bottom of a page, with no text underneath it. You need to check a printed version as you will see features you don't spot on a screen.

12.5 Mathematical Equations

Most science theses or dissertations will include some mathematical equations. In some cases these will be straightforward equations that can be written just using normal characters and a few special symbols, which are easily produced using most word processing packages. In other cases, particularly in subjects like physics and mathematics, there will be many complex equations that cannot be presented using standard characters, and it will be necessary to make use of more specialist software of some sort.

---

*Like this footnote.

Individual mathematical symbols and short equations can be included in the text, so for example, you can refer to a field with a length l and a width w and write the equation for the area $A=lw$ in the text. But even in this simple case note that you need to italicise the symbols to make it clear that they are representing mathematical quantities. If you just write A=lw it does not look professional.

Longer equations should be written as display equations taking up a whole line. For example the equation for the solution of a quadratic equation $ax^2 + bx + c = 0$ can be written as

$$x = \frac{-b \pm \sqrt{b^2 - 4ac}}{2a}$$

and the best way to do this is by using a specialist equation add-in to a program like Word or OpenOffice. Alternatively, if there are lots of equations like this in your thesis you should consider using LaTeX, which has enormous flexibility. The alternative of trying to write the equation using standard symbols, for example,

$$x = [-b \pm \sqrt{(b^2 - 4ac)}]/2a$$

is mathematically correct (apart from the failure to use italic font) but looks very messy and is harder to read than a properly displayed equation. Other more complex equations such as the equation for a Fourier series

$$f(x) = a_0 + \sum_{n=1}^{\infty} \left(a_n \cos\frac{n\pi x}{L} + b_n \sin\frac{n\pi x}{L} \right)$$

or for matrix manipulation

$$\begin{pmatrix} x' \\ y' \\ z' \end{pmatrix} = \begin{pmatrix} a_1 & b_1 & c_1 \\ a_2 & b_2 & c_2 \\ a_3 & b_3 & c_3 \end{pmatrix} \begin{pmatrix} x \\ y \\ z \end{pmatrix}$$

are very hard to write without using specialist functions.

If you look carefully at any scientific textbook or at journal papers you will see that there are a number of conventions for writing mathematical symbols and equations that should be followed. You are probably not aware of them normally but if you don't follow those conventions your work will not look professional and it will be harder to follow.

Here are some of the more important conventions:

- Use italics for mathematical symbols such as x and y – both in the text and also in display equations
- Note, however, that numbers (0, 1, 3.14159) and mathematical functions such as sin, cos, log and exp should *not* be in italics
- Similarly, unit symbols such as m, J and Da should *not* be in italics
- Chemical elements should be written without capitals and *not* in italics (helium, oxygen, magnesium) but their abbreviations do have a capital (He, O, Mg)
- Use Greek letters (α, β, γ) where appropriate
- Display equations should be centred and numbered if you want to refer back to them later on.

The appearance of equations in your thesis makes a significant difference to the first impression the reader gets when they open it up. It pays to make sure that the equations are presented well. Even though preparing the first few is time-consuming, it soon gets easier because you become more efficient at using the software and also because you can save time by copying and pasting from equations you have already written.

Common Mistakes

- Inconsistent layout
- Failing to separate paragraphs and other sections clearly
- Excessive numbering of sub-sub-sub-headings
- Leaving a heading as the last line of a page
- Not checking a hard copy of a thesis, but just checking the on-screen version.

Key Points

- Use a clear, easy to read font, such as 12 point Times
- Use one-and-a-half or double line spacing (and check your local rules for thesis layout)
- Leave enough space on the left margin for binding
- Check the final layout on a printed version of your manuscript.

Endnote

[i] Like this endnote.

Chapter 13

Numbers, Errors and Statistics

13.1 Writing Numbers and Quantities

Numbers

In any scientific document, including a thesis, it is important to be careful about how numerical quantities are written, because you want to avoid any possibility of confusion, or, worse, presenting wrong information.

Always include a digit before the decimal point (write 0.034, not .034). If there is a decimal point, always include a digit after the decimal point (write 34.0, not 34.). Note that elsewhere in Europe the normal symbol used for the decimal point is a comma, so this can easily cause confusion.

Figures should generally be grouped in threes for long numbers; technically, the correct separator of the groups for scientific writing is a space, but in documents written in English a comma is also usually acceptable: write 3.141 592 654, or 3.141,592,654, but not 3.141592654. Similarly, write 64 000 000 or 64,000,000 but not 64000000.

Quantities

For writing physical quantities such as lengths, temperatures or concentrations, it is important to use SI units. The *Système international d'unités* (international system of units) is the modern version of the metric system of units and was formally agreed in 1960. It defines seven *base units*:

- For length: the metre (m)
- For time: the second (s)
- For mass: the kilogram (kg)
- For thermodynamic temperature: the kelvin (K)
- For electrical current: the ampere (A)
- For luminous intensity: the candela (cd)
- For amount of substance: the mole (mol).

There are also many derived units, including the radian (rad) for angle and the steradian (sr) for solid angle. Others are named after individuals such as the newton (N), joule (J), hertz (Hz), volt (V) or pascal (Pa). Note that the symbols are capitalised when the unit is named after a person, but all the unit names are written in lower case letters, even if named after a person (for more details, see *Appendix 7: SI Units*).

When writing quantities, there are well-defined conventions given in the SI guidelines. The reason for these conventions is to ensure there is no ambiguity in the meaning of a written quantity. Although some of the rules may seem very detailed and unnecessarily fussy, they are worth following carefully to avoid any confusion. Beware of the examiner who requires you to change every incorrect unit! Some of the most important points to remember are:

- Always use the correct abbreviations for the units (for example, use s not sec for seconds)
- If you use kelvin (K) for a temperature, do not use it together with the degree symbol (°)
- Do not italicise units, for instance write 5 mm not 5 *mm*: note that if you include the unit inside an equation in some word processing packages it will automatically be italicised because the software thinks it's a mathematical symbol, so you may need to override this or make sure to keep the units separate from the equation

- Do not add 's' on the end of the unit to make it a plural: 2 kg is correct, not 2 kgs
- Use prefixes such as mega (M), kilo (k), milli (m) and micro (μ) as appropriate (a full list is given in *Appendix 7: SI Units*) – note that there is no space between the prefix and the unit, for example, 3 kV not 3 k V
- Separate the value from the unit by a 'non-breaking' space so that you do not end up with the value and the unit on different lines of text – refer to the 'Help' pages of your word processor to find out how to do this
- Although percent (%) is not formally an SI unit, it's generally best, but not essential, in scientific writing to put a non-breaking space between the number and the % sign, and this is the recommendation in the SI documentation, for example 15.4 % not 15.4%
- You can combine units as necessary, for example, for velocity you can write m/s, m s^{-1} or m·s^{-1} and for density kg m^{-3}, kg·m^{-3} or kg/m^3
- A quantity (for example, force) that has a derived unit (newton) can we written either in terms of the base units (kg m s^{-2}) or the derived unit (N) – these are equivalent
- If you are writing down the dimensions of an object you need to show units for both quantities, so you might write 6.4 m × 5.9 m or (6.4 × 5.9) m but not 6.4 × 5.9 m as the unit for the value 6.4 is then ambiguous.

If you do need to use a non-standard unit – for example, if your subject has a different convention – make sure that you give a clear definition of your units so that someone who is not familiar with the field can understand what you have written. For example, it is still common to find vacuum and pressure gauges calibrated in the non-SI unit Torr (where atmospheric pressure, defined as 1 bar ≈ 10^5 Pa, is equal to 760 Torr). So although the correct unit to use for pressure is normally Pa, in some cases it makes sense to quote pressures in Torr, so long as a definition is given.

A final point: make sure that every numerical quantity that you give in your thesis includes the appropriate unit. It's an obvious point but a statement like 'The length was 6.2' really is meaningless without the unit.

13.2 Errors and Uncertainties

All physical measurements have some degree of uncertainty associated with them as it is never possible to make a perfect measurement. Although uncertainties are commonly referred to as experimental errors, this does not mean that the person making the measurement has made a mistake! It's an unfortunate use of the word 'error' and it simply expresses the fact that the measurement has a limited accuracy.

Random errors and systematic errors

There are different types of errors (or uncertainties) in experimental measurements and they divide mainly into two categories: systematic errors and random (or statistical) errors. If you are unsure of the classification of errors or how to deal with them, refer to one of the many good textbooks on this subject as soon as possible and certainly before you start analysing your results. You don't want to have to go back to the start if you find that you have misunderstood the significance of an experimental error.

Systematic errors are ones that affect each measurement the same way (for example, if an instrument has an incorrect calibration, all measurements will be wrong by the same amount) whereas random errors are those that affect each individual measurement differently (for example, if a signal is noisy, each time you take a measurement you will get a slightly different answer).

The good news is that random errors can be reduced by making more measurements; essentially, you are just averaging out the errors. Normally, if the random error of a single measurement is σ, the random error of N measurements will be $\sigma/\sqrt{N}$. This means that you can reduce your random error by a factor of about 3 by taking 10 measurements and finding the mean value; however, if you want to reduce it by another factor of 3 you will have to take around 100 measurements, so this is a game of diminishing returns.

There is an important point here about nomenclature for random errors. The error of a single measurement is described mathematically using the *standard deviation*, which measures the typical amount by which a single measurement deviates from the 'true value' of the quantity. If lots of measurements are combined, the standard deviation of the measurements remains the same, but the accuracy with which the true value can be estimated improves by the factor mentioned above. The new uncertainty is called the *standard error of the mean*, which gets smaller by a factor of $\sqrt{N}$ as the number of measurements (N) increases. Different fields have different conventions for how you write the final error value for a quantity you have measured but in the physical sciences you will generally need to give the standard error of the mean – check with your supervisor about your field.

Most random errors follow a 'normal distribution', meaning that if you were to plot a histogram of a very large number of measurements of one quantity you would obtain a graph that always has the same shape (often described as the bell curve). The mathematical rules that are used to manipulate error values follow from the assumption that the errors follow this distribution. Any book on errors will show how these very powerful results are obtained.

Error calculations

Before you embark on writing the numerical parts of your thesis, you should check that you know and understand the rules for dealing with errors. This will ensure that you are able to extract the best possible values from your data (which will always be limited as no one ever makes a perfect set of measurements). It will also ensure that you will not make any unjustified claims for the accuracy of your data – all examiners will have come across a thesis in which the errors in the final results were not correctly calculated from the measurements – and so you will avoid being asked by the examiners to make corrections to the thesis.

Here are two important things you need to think about when dealing with errors. First, try to understand what the errors are in your experiment. Do you know which are systematic and which are random? Can you estimate the values of your systematic errors? Can you eliminate or measure your systematic errors by making any additional measurements? Can you reduce your random errors by making repeated measurements?

Second, make sure you know how to combine errors from different quantities. Often your final result is calculated from measurements of different parameters, each of which may have errors. For instance, if you want to determine the density of a liquid by measuring its volume and mass, do you know how to find the error in the density, given the errors in the volume and mass? If not, look it up in a book on errors.

How to write errors

Write down your results and their errors correctly. This ensures that you express precisely how well you have measured the given values. Normally errors are written in the form $a \pm b$ with the meaning that the true value of a probably lies between the values $a - b$ and $a + b$. But what do we mean by 'probably' here? In many cases you can only estimate the error from a reasonable assessment of the accuracy of the measurement. But if the error has been calculated using a statistical analysis of a set of measurements, and b is the standard error of the measurements, then the probability that the true value lies in this range is approximately 68 % – this follows from the definition of the standard error and you can find the mathematical details in any textbook on measurements and errors. If you extend the range by a factor of two, then the probability is 95 % for the quantity to lie within the range $a - 2b$ to $a + 2b$. A range of plus or minus three standard errors gives a probability of 99.7 %. The convention is normally that you quote a range of plus or minus one standard error. Sometimes you may want to quote a value giving a much higher confidence that the range $a - b$ to $a + b$ includes the true value, in which case you may use an error value (b) of twice the standard error, but in this case you *must* say that this is what the value b represents.

Here then are a set of guidelines for writing down your results with their errors:

- Always give the measured value and the error in the same units
- Because the unit applies to both the measurement and the error, use brackets or write the unit twice to avoid any ambiguity, i.e. (293.1 ± 0.4) K or 293.1 K ± 0.4 K, but not 293.1 ± 0.4 K because this

does not specify unambiguously the unit for the value 293.1, although many people do not follow this rule in practice .

- Use an appropriate number of significant figures, to reflect how well you know the value: if you write $g = (9.80665 \pm 0.16342)$ m/s, the last three digits on both value and error gives no useful information and should not be there; this result should be written $g = (9.81 \pm 0.16)$ m/s
- Always quote the error to the same number of decimal places as the value itself – if you write (6.1 ± 3) mm or (9 ± 2.7) s the meaning is ambiguous, but (3.4 ± 0.5) mg is unambiguous and correct.

Alternatively, you could use a percentage value to express an uncertainty, for instance ± 5 %. This is sometimes a more convenient way to express a result. For more precise measurements, you may quote an error as 'ppm' (parts per million, i.e. units of 10^{-6}), 'ppb' (parts per billion, 10^{-9}) or even 'ppt' (parts per trillion, 10^{-12}). In some fields, these are the standard ways of writing uncertainties.

Another alternative, common in some fields, is to write an error using brackets, where the number in brackets is the error in the final digit(s) of the value. So, for example, 7.436 (12) s means the same as (7.436 ± 0.012) s.

13.3 Statistics

It goes without saying that it's really important to get any statistical calculations in your thesis right. We have come across several instances where a thesis has contained conclusions that were not justified statistically by the data presented, resulting in the candidate having to rewrite parts of the thesis. So check your use of statistics very carefully and ask people for advice if necessary.

We won't go into any detail here about the statistical techniques that you will need as these depend on your field. All we do here is to discuss some general principles about the use of statistics in presenting the conclusions from your work.

What are statistics for?

We use statistics to extract as much information as we can from a set of data. In the real world we can never gather as many data as we would like to. This means that we need to use special techniques to make sure that the conclusions we draw from the data can be justified. Normally this will mean that our conclusions are stated in such a way as to qualify their validity: for instance we might say 'The wavelength of the light is (683.7 ± 0.3) nm' or we might write 'There is a significant difference in the scores between group A and group B ($p <$ 0.001)'. Here the expression in brackets means that based on the data, the probability of seeing a difference in scores at least as big as this purely by chance is less than 0.1%. In both cases we are expressing the limit in our knowledge. This tells the reader how reliable our conclusions are.

Statistical tests

There is a number of statistical tests that use the given data to quantify the likelihood that a statement is true. Different tests are used for different types of data and you should refer to a statistics textbook for details of which test to use in which situation. Most statistical tests you are likely to use are based on a number of assumptions, which may include a normal distribution of the data, homogeneity of the variance, and independence of the samples; it is important you think about all these factors before you start your analysis.

The two most common statistical tests are given below.

χ^2 *test (chi-square test)*

The χ^2 test assesses the degree to which the statistical fluctuations in your data are consistent with a hypothesis you are making. This can be used for categorical data, for example 'dead' or 'not dead', or 'red' or 'blue'.

An example might be an assessment of whether being vegetarian affects your ability to wiggle your ears. You might question lots of people and divide them into four groups according to whether they are or are not vegetarian and whether they can or cannot wiggle their ears. The χ^2 test could then be used to tell you whether you would be justified in hypothesising that one was correlated with the other. Here the χ^2 test is testing *independence* of data.

A different example would be a test of whether a set of measurements fitted a theoretical model. For instance, if you measured the length of a rod at a set of different temperatures, you could plot the length as a function of temperature and try to fit a straight line to the data. The χ^2 test tells you whether the scatter of points from the straight line is consistent with their individual errors. If they are not, it means that your model (the hypothesis that the length changes linearly with temperature) is not justified by the data. Here the χ^2 test is testing *goodness of fit* to the data.

In order to interpret the results of the χ^2 test, you need to know about the number of *degrees of freedom* in your data. For discrete data, this is a measurement of your sample size relative to the number of variables you are studying. For continuous data it is a measurement of how many independent variables there are. Tables are then used to find the probability of finding a particular value of χ^2 as a result of random fluctuations. Determining the number of degrees of freedom can be complicated and it depends on the type of data that you have, so we will not go into this here. Suffice it to say that you need to make sure that you have assessed the number of degrees of freedom correctly, as otherwise your conclusions may not be valid.

Student's t-test[1]

This test is used to assess whether there is a significant difference between two results, and it can only be used on continuous data (for instance, the average scores in a test carried out on two groups of

---

[1] Note that this test is named after 'Student', which was the pen name taken by William Gosset. He worked at the Guinness factory in Dublin in the early 1900s but needed to hide his identity because the use of statistics to monitor beer production was commercially sensitive information.

people). The t-statistic is a measure of how different the two scores are in relation to their statistical uncertainties. High t indicates a more significant difference than low t. Special tables are then used to express the meaning of that value of t in terms of the probability that that difference could be obtained by chance.

Statistics software packages

In many cases, with complex or extensive datasets, it is necessary to use specialist software packages to analyse the data. There are many such packages available. Some are free and easily available online while others require a registration fee. Some other software (such as spreadsheet programs and graph-plotting programs) will include the most important statistical functions as well. One problem is that these packages will only give you reliable results if you use them appropriately, so having a powerful statistics package available to you does not mean that you can avoid reading up on which test is appropriate and how to interpret the results. Remember the old adage: 'Garbage in, garbage out!'

A particular example of this is the use of a straight line fit. Often the error on the slope (or gradient) obtained from a straight line fit is surprisingly small, and the temptation is to use this in further calculations without further thought ... but you need to remember that this error simply tells you how accurately the slope and intercept of the straight line can be calculated from the data. It does not tell you how reliable the data are, how well the straight line fits the data (see Figure 13.1) or whether the straight line is the appropriate model for the data (see Figure 13.2). So you need to consider carefully what really determines the correct value you should use for the error in the slope. Note that the data presented in Figures 13.1 and 13.2 are such that the best fit intercept and slope, and the standard error of the slope, are all the same for both plots. These examples show that you have to make sure you check a plot of the raw data before you make use of any output from a curve-fitting program.

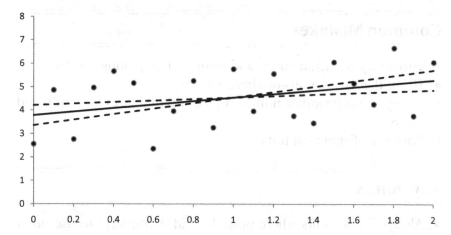

Figure 13.1 Straight line fit to some noisy data. The best straight line fit (solid line) is shown together with lines having slopes one standard error greater and less than the fitted value (dotted lines). The error value of the slope coming out of the fit is lower than might be expected and needs to be used with care.

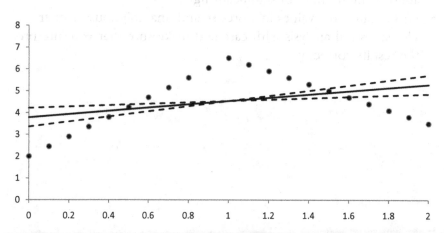

Figure 13.2 Straight line fit to some data that does not follow a straight line. The best straight line fit (solid line) is shown together with lines having slopes one standard error greater and less than the fitted value (dotted lines). The error value of the slope coming out of the fit is lower than might be expected and needs to be used with care. A single straight line is clearly a very poor model for these data. Nevertheless, the error value of the slope coming out of the fit is much smaller than the range of slopes present in the data. Clearly, it would be wrong to read any significance into the values coming out of the straight line fit in this case.

Common Mistakes

- Forgetting to include units and errors when quoting results
- Using the wrong units or abbreviations
- Using an inappropriate number of significant figures for results and errors
- Poor use of statistical tests.

Key Points

- Always use SI units where possible, and if you have to use other units, give a clear definition
- Distinguish between systematic and random errors and use the correct procedures for calculating them
- Always quote errors for experimental measurements, using an appropriate number of significant figures
- Write your error values in a precise and unambiguous manner
- Use statistical analysis with care and make sure that you interpret the results correctly.

Chapter 14

Figures and Tables

Figures divide broadly into two types: illustrations and graphs. We use illustrations to give evidence that supports a statement in the text, or to help explain or clarify a written description (for example, a map, a photograph of a plant or a diagram of an experimental setup). We use graphs to display data in a pictorial or graphical format (for example, a plot showing rates of vegetative propagation in plants, a histogram of radioactive decay events or voltage signal as a function of time). We use tables to present large amounts of numerical or qualitative data in an easily readable form.

Before you start to make your figures and tables, read your thesis plan and make a list of what to include. This will help you to get a good balance between the data coming from different parts of your project and to include everything necessary to show what you've achieved. Remember that you are the expert, and things that are obvious to you may not be obvious to your reader, so don't forget to include diagrams or other illustrations to help your reader understand your experiments and data.

14.1 Basic Principles

Figures and tables need to be clearly organised. Make sure you know what information you are trying to convey, and plan your figure or table so that you do so efficiently without including redundant ideas or data. Do not try to cram in as much information as you can into one figure or table, as this will make it difficult to follow. Do not repeat information from your figures and tables in the text: simply refer in the text to the point you want to highlight from the figure or table. Remember to make clear where the data that have gone into a figure or table came from.

Figures and tables are created in a number of ways, depending on the type of data, and the computing facilities available to you. Before you start to create a figure or table, draw a rough draft of the contents and layout. This will help focus your ideas and give you an idea of the relative size of the components and the complexity of the figure or table. Making illustrations can be very time-consuming so you do not want to waste time revising them. Although all figures and tables must be referred to in the text, they should not depend on the text to be understandable. When planning figures or tables there are four considerations to take into account:

1. The graphical display (does it look good and show what it ought to?)
2. Annotation (to explain features in the figure or table)
3. The figure or table title
4. The figure or table legend (a short piece of text that tells the reader what they are looking at).

Figure 14.1 shows the same data treated in three different ways. One is good, one is bad, and one is ugly and bad. Figure 14.1(c) is good and complete because it includes all the above four features.

Size

The simpler the figure or table is, the smaller you can make it; but it has to be easily readable and look good on the page, so do not make it too small. Remember your examiner may be over 50 and not able to see as well as they used to – so try to use a 10 or 12 point font and certainly never use anything smaller than an 8 point font. With a more complex figure or table you will need to make it large enough to present the information clearly. Try not to make your figures or tables larger than one side of a page or you will have to fold them over to fit in your manuscript; but if you really do need to have such a large figure in order to show the information clearly, then do so.

Figure 1.1

(a)

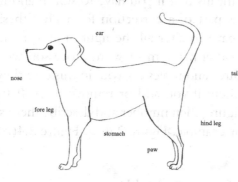

Figure 1.1 A dog.

(b)

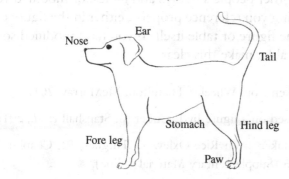

Figure 1.1 **A dog.** The external anatomy of a dog
is shown; the stomach is an internal organ.

(c)

Figure 14.1 **A figure.** (a) Graphical display, without annotation, a legend or title;
(b) with badly placed labels that are too small and so it is unclear what is being
annotated; (c) the same figure is much clearer with annotation, a title and legend.

You can put more than one figure or table on a page, but leave a large enough margin around each one to clearly separate them.

Numbering

The easiest way to number your figures and tables (both for you and the reader) is by giving the chapter number and then an individual figure or table number; for example, in Chapter 2, the figures (including any graphs) would be numbered Figure 2.1, Figure 2.2, Figure 2.3 and so on; the tables would be numbered Table 2.1, Table 2.2, Table 2.3 and so on. This means that if you have to slot in another figure at the last minute, or as part of a correction for a PhD thesis, for example, then you don't have to alter all the figure numbers in the remaining chapters (on the other hand, many word processors will now automatically number figures and tables for you: in this case they will automatically renumber them if you add or remove one). If there are two or more parts in a figure, then number and describe them separately in the figure legend, for example, Figure 2.4(a), Figure 2.4(b), etc.

Other people's figures and tables

It is fine to use other people's figures and tables, or modified versions of them, providing you reference properly, either in the figure or table legend, or on the figure or table itself. If you have modified someone else's figure or table, make this clear.

Fig. 1.5 is taken from Wheeler, Hamilton, McMurray, 2017.

Fig. 3.6 is based on a figure by I. Cutler (in Stanshall *et al.,* 2016).

Table 8.5 is taken from Rice-Oxley, T., Hughes, R., Chaplin, T., Quin, J. 2015 (Supplementary Material online).

If your thesis is going to be put on an open access website by your university (or by you!), you may also need to get permission to reproduce figures, even for those taken from your own publications (see *Chapter 11: Other People's Work*).

14.2 Figures: Illustrations

Illustrations need to be as carefully planned as your text. Put in figures that help clarify any ideas or information you are presenting. These might be diagrams summarising our current state of knowledge of the system you are researching, or they could be flow diagrams describing the strategy you took, or they could be photographs of apparatus or geological features or insects' wings – whatever helps explain your text.

Creating figures

Figures can be prepared from a number of different sources and may be:

- Figures drawn by hand
- Original photographs and scanned images
- Schematic diagrams, flow charts, layouts, etc., created in a graphics or drawing program.

Figures drawn by hand

Nowadays it is extremely unlikely that you will need to draw figures by hand, but if you do, there are a few points to note. Unless you are a very gifted artist, it is best to keep these figures simple. If you are drawing by hand, go to your nearest art shop and invest in a set of draftsman's pens. These give clear, even, dark lines that are ideal for producing high-quality diagrams. Do not use a pencil or a pen that is likely to smudge.

Original photographs and scanned images

Photographs should be clear and in focus, and only include information that is of use to your reader. If, for example, you are photographing a piece of equipment, do so against a neutral background and do not have unnecessary details intruding into your figure. Similarly, if you have a small image surrounded by a sea of background, get rid of the background as it does not tell the reader anything. Use a digital camera if you can, but if this is not possible use a good-quality scanner to convert your photograph into a digital format. Clean up your photograph with a suitable editing package if necessary. If your photograph does not clearly illustrate what you want it to, consider using it alongside a line drawing to highlight the important details.

Many instruments (for example, microscopes and telescopes) now produce visual images that are directly read into a computer. Other digital images (including line drawings and traces on paper from recording instruments) can be created by using a scanner. Once they are in a digital format you can use drawing or graphics programs to arrange and annotate the images so they show the important information clearly.

All sorts of images can be manipulated with the aid of a computer program, and this is often used to enhance images – but of course they should not be used to alter data. Take care manipulating images as there is a danger of giving a false impression of the experiment, which could be construed as fraud!

Figures created in a graphics or drawing program

There are many drawing and graphics programs available and most are fairly easy to use, at least to produce simple images (see *Chapter 2: Getting Organised*). It is worth spending a little time playing with these programs and learning how they work; they will allow you to produce clear and neat images that can easily be edited and rearranged if necessary.

Some diagrams consist of line drawings showing the layout of the components of an experimental arrangement (for example, Figure 14.2). Try to arrange the different elements in such diagrams in a logical way, and make sure the lettering is large enough to be easily read. Give an explanation in the legend to clarify any feature on the diagram that is not obvious.

14.3 Annotating Figures

Annotation simply means adding explanatory labels – such as arrows, numbers or letters – onto the figure to indicate special features that you refer to in the figure legend or text. Most figures need to be annotated in some way. When adding annotations make sure they are large enough to be easily seen, but not so large they overwhelm the image. Letters and numbers in 10 or 12 point font are usually all right; never

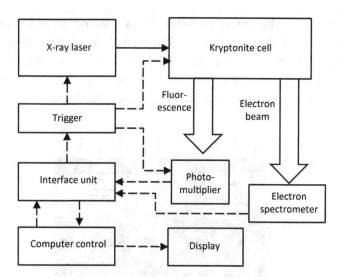

Figure 14.2   An example of a schematic diagram of a set of apparatus. Solid lines indicate the laser beam and dashed lines indicate the flow of electrical signals.

go below an 8 point font. Do not italicise the letters or put them in bold, unless that is the correct convention; for example, gene symbols are italicised. You can use Arabic (1, 2, 3, 4), or Roman (i, ii, iii, iv) numerals, and either capital (A, B, C), or lower case (a, b, c) letters. Lower case letters usually look neater. Keep your annotations clear and simple and place labels as close as possible to the objects they refer to. Do not overdo your annotations or you may clutter the figure.

If your data are in a drawing or graphics file on your computer, then use the program to annotate the figure. Choose a font that is easy to read and use it consistently throughout your thesis or dissertation.

Labelling items in a figure

If you have to label a large number of items in one figure, your labels can be vertical, rather than horizontal. It is usually clearer if you give the name of an item rather than a letter key that needs to be explained in the legend. Generally speaking, keys are a nuisance for the reader, but if you do not have room to write the name, then choose your letter key sensibly. For example, if you are electrophoresing RNA samples from different tissues, labelling them B, H, L, for brain, heart, leg, will make it easier for the reader to follow what is going on in the diagram than if you label them A, B, C. When labelling figures with letters, make sure that all the letters align if they need to and are all on the same horizontal line and the same vertical line as appropriate. Otherwise the figure will look messy and unprofessional (see Figure 14.3).

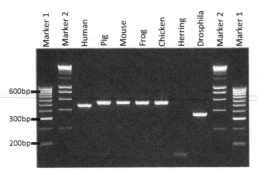

Fig. 3.6   Results of PCR to detect *Beegee* exon 1 from DNA samples.

(a)

Fig. 3.6   Results of PCR to detect *Beegee* exon 1 from DNA samples.
M1, M2, molecular weight markers; Hu, human; P, pig;
M, mouse; F, frog; C, chicken; He, herring; D, Drosophila.

(b)

Figure 14.3   A figure containing many samples. (a) Each sample is labelled clearly with vertical annotation, and a scale is included. The annotations align vertically and horizontally. The figure is a good size so that it is clearly readable but does not take up unnecessary room on the page; (b) the samples are labelled with a key that clearly indicates which is which and, again, this is a clear, readable figure; (c) the samples are labelled with a key that is non-obvious, the letters of the key do not align on the figure and they face the wrong way. There is no indication of size or scale, one lane is not labelled, and at least two-thirds of the figure is blank space. Also, in this particular example, the gene name (*Beegee*) should be italicised. Finally, the image is ENORMOUS and takes up unnecessary room on the page.

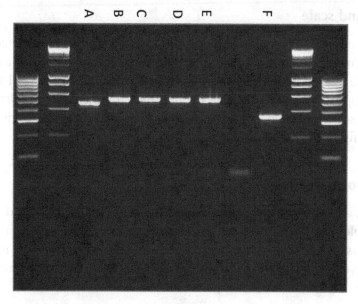

Fig. 3.6   Results of PCR to detect Beegee exon 1 from DNA samples.
A human; B pig; C mouse; D frog; E chicken; F Drosophila.

(c)

Figure 14.3   Continued.

Always present your control or 'normal' sample first, so the reader looks at this first and then can compare the next values to this. If a number of related samples appear on one figure, put them in a logical order. If the same samples appear on many figures, keep them in the same, logical order in each figure. If you show similar data in more than one figure, try and keep to the same scale within each figure, and the same size for each figure. Help the reader by standardising your displays as much as possible.

Remember to label all the components you have included in the figure. Unlabelled data will make the examiner think you have forgotten something. If you have to incorporate data that are irrelevant to your text – for example someone else's samples, or data from an experiment that didn't work, but was run at the same time as your samples – then put an asterisk and explain the traces are not relevant in the figure legend. Let the examiner know that you have thought about what you are showing them.

Size and scale

Tell the reader as much as is necessary about the size of the objects in your image. Either label each object with its size, or provide a marker or scale against which each object can be compared. In Figure 14.3(c) it is impossible to tell the size of the DNA fragments in the diagram as no scale has been provided. Such information has to be given in every figure for which is it relevant.

Composite figures

If you are labelling different parts of the same figure, a, b, and c, for example, put each label in the same place in relation to the image, such as in the top left-hand corner or centred below the image. However you choose to annotate your figures, be consistent (see Figure 14.4).

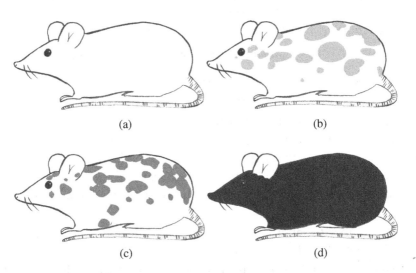

(a) (b)

(c) (d)

Fig. 4.2 **Different types of mouse.** (a) A white mouse; (b) a light-grey spotty mouse; (c) a dark-grey spotty mouse; (d) a black mouse with white ears.

Figure 14.4 A figure made up of different components. Each part of the figure is labelled, (a), (b), (c), (d), and the labels are in a clearly readable font and are positioned in the same place in each section.

Figure titles

The figure title provides a heading for your figure, which will be listed in the Table of Figures at the front of your manuscript (see *Chapter 10: The Other Bits*). Like all other headings it should be concise and descriptive: give the figure number and just one short phrase telling the reader what is there. The figure number can be written either as *'Figure'* or *'Fig.'*. Put the title in bold if you wish to. If you have a diagram you could write *'Figure 1.1 Diagram of ... ',* if it is a photograph you could put *'Fig. 5.3 Photograph of ... '.* Most people prefer to simply state the subject matter, *'Figure 9.6 Mass spectrogram of sample A', 'Figure 8.3 Distribution of landfish in Kellington', 'Fig. 2.5 Time dependence of absorption coefficient'.*

However you write your titles, be consistent. In most theses and dissertations the figure title is placed below the figure and is followed by the figure legend.

Figure legends

Figure legends give the reader a more detailed description of what is in the figure. They explain the contents of the figure and where any data came from. Use the figure legend to describe interesting features that you have annotated. If you have had to label multiple samples with a key, then use the figure legend to explain what the key represents. Legends also allow you to cite references that are relevant to the figure. There are no rules about the size of figure legends, but very long legends are difficult to read. It is not necessary to have a legend if the title says all that is required.

If you are referring to multiple items as in Figure 14.4, place the '(a)' in front of the explanation, as we have done in the legend. Do not put it afterwards (i.e. A white mouse (a); a light-grey spotty mouse (b) ...) because this is extremely confusing to read.

Do not reiterate the contents of a figure legend in your text: the two are telling the reader different things. The figure legend is a detailed description of the figure, while the text discusses the important highlights. Remember to spell-check figure legends, and make sure they actually describe what is in the figure.

Black and white or colour?

Generally there are no rules regarding the use of colour in theses and dissertations. If you can use colour (because you have access to a colour printer), use the colours thoughtfully. Colours can make a figure clearer, but if you go over the top it will look more like a psychedelic album cover than a scientific figure and will be difficult to follow. Make sure related features have similar colours: for example, a graph of the use of hair conditioner by the super-groups Deep Purple, Black Sabbath, Led Zeppelin, Procul Harum and Thin Lizzy might well show the data in shades of ... well, purple, whereas the same graph might show the comparable data from Bananarama, Blondie and Right Said Fred in shades of yellow. Possibly.

Don't let one colour dominate, unless you want it to. Putting all your data into pastel shades means that the single line in scarlet will dominate over the rest of your data.

Also bear in mind that someone may well print out a copy of your thesis in black and white, and you will still want the figures to be clear and intelligible in that case too. Therefore try to use colours that will not look the same when printed in monochrome. Consider using different line formats (e.g. dotted or dashed) or different types of shading in addition to colour if necessary.

14.4 Figures: Graphs

A graph is a type of figure that displays a summary of numerical data in an easily understood manner. Graphs get the message across more immediately than if the reader has to scan through a table of numbers, and they are extremely useful for showing the distribution of data, highlighting trends and summarising relationships. Graphs present quantitative data in many different ways including conventional x–y plots, scatter plots, histograms, pie charts and contour plots. Make sure that you use the one that is most appropriate and conveys the information in the most efficient manner.

Graphs are essential for showing your primary data to your examiners. You do not need to show every piece of data you have produced during the course of your project, but you do need to let the examiners see your important results, those that enabled you to draw a conclusion, and perhaps led you to the next experiment.

You need to show in your figures the experimental data that you are discussing in the text. If you have carried out a set of similar experiments and have a lot of examples of the same kind of data, choose a representative set of graphs to illustrate what you have done. You can put the rest of the data in an appendix if necessary.

Many types of data are gathered automatically on a computer (such as measurements of optical density or a frequency or mouse locomotion) or may be generated by the computer (outputs of calculations or simulations). Other data may have been entered by hand from measurements. In most cases you will probably need to carry out some analysis of the data and plot the results.

There are many easy-to-use computer programs that draw beautiful graphs from tables of data. It is worth spending time learning how to use these programs. Refer to the online help or manual for further information about producing different types of graph (see *Chapter 2: Getting Organised*).

Here are some guidelines for creating the simplest types of line graphs, scatter graphs, histograms, bar charts, pie charts and contour plots. If you are unsure which type to use for your analysis, consult your tutor or supervisor. Many readers of this guide will be producing considerably more complex types of graphs, such as those with non-Cartesian coordinates or multiple axes; your supervisor is your best guide for help in creating these graphs, but the basic guidelines are the same.

Creating graphs

It is now very easy to generate a simple graph from spreadsheets, but for scientific writing such as a dissertation or thesis this will generally not give you an acceptable graph. Figure 14.5 shows an example of the default scatter plot from one spreadsheet program for a set of data.

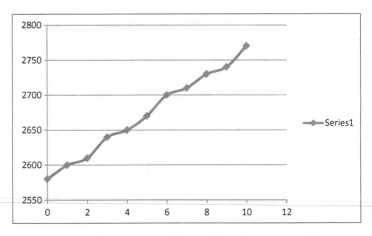

Figure 14.5 **The default scatter plot produced by a spreadsheet program.**

There are many things wrong with the plot in Figure 14.5. For example:

- The x- and y-axes are not labelled to say what they represent
- The y-axis labels have too many digits
- There is a superfluous 'Series 1' annotation
- The symbols used for the data points are too large and imprecise
- The curved line joining the points serves no purpose
- The x-axis scale goes to 12 but the data points stop at 10
- There are unnecessary grid lines and a box around the figure.

Figure 14.6 shows an improved plot of the same data, in which all these points have been dealt with and the result is a much more professional-looking graph. This is the sort of figure that you would expect to see in a published journal article and you should aim to show plots of this quality in your thesis.

Generally, values that are chosen in advance in an experiment, such as time intervals, are plotted on the horizontal, x-axis; these are known as independent variables or control variables. Variables that are measured at these points are plotted on the vertical, y-axis; these are known as the dependent variables or response variables.

You do not need to show the origin (i.e. the point where x and y are both equal to zero), providing you clearly label the value of each

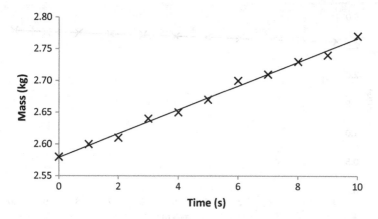

Figure 14.6   A plot of the same data as in Figure 14.5. Some improvements have been made to the formatting. A straight line fit to the data has also been included.

axis where they intersect. Indeed in many cases it is best *not* to include the origin in the graph (see Figure 14.7).

For some graphs it is appropriate to include error bars with your data points. Ask your supervisor for guidance if you are not sure whether to include them. If you do include error bars, make it clear in the figure legend exactly what they represent.

Annotating graphs

Just like any other figure, graphs need to be annotated so the reader can interpret them, but try to keep annotations to a minimum, otherwise graphs tend to look very cluttered. You should:

Label what is represented by each axis. Use simple labels of just a few words for the label, such as 'Time' or 'Frequency', and use standard abbreviations. Do not capitalise all letters in a word: use italics or use bold lettering for your labels as capitals tend to look crude or fussy.

Label each axis with a unit of measurement. Use standard abbreviations and SI units ('m', 'kg', 'cd', etc.) wherever possible. If this is impossible provide an explanation in the legend to your graph. You can write 'Time (s)' or 'Time/s' but the first is more commonly used.

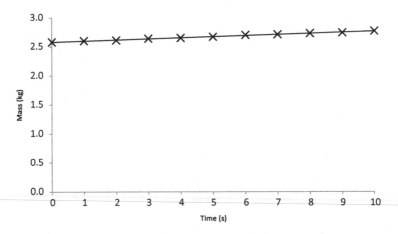

Figure 14.7 Choose the appropriate scale for presenting your data. This plot is also for the same data as in Figure 14.5. The scale used here is not a sensible choice as it prevents the reader seeing the behaviour of the data clearly. Also the labels are too small for them to be read easily (at least by the authors of this book).

Mark the divisions of your scale along each axis, and mark in sensible multiples. Provide a scale along each axis. You do not need to mark every unit: often it is neater and as clear to mark the units in multiples of 5 or 10.

Types of graph

Line graphs

Line graphs present a precise relationship between two (or more) quantities, one plotted along the horizontal, *x*-axis and the other along the vertical, *y*-axis. This might often be the case for the results arising from a theoretical model of a system. The information is presented as a line rather than points as it is a continuous relationship and does not depend on individual data points. Figure 14.8 gives an example of such a graph.

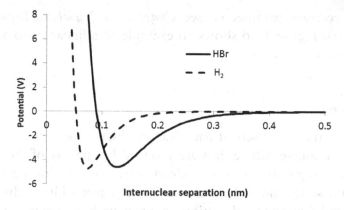

Figure 14.8 Plot of the Morse potential for HBr and H_2.

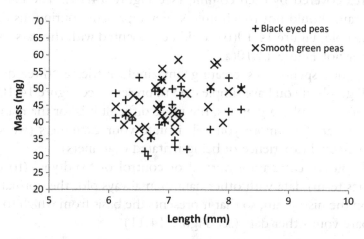

Figure 14.9 A scatter graph. The two sets of data are distinguished by using different symbols.

Scatter graphs

Scatter graphs present the relationship between two different types of information plotted on a horizontal, *x*, and vertical, *y*, axis. You simply plot the point at which the values meet, to get an idea of the overall distribution of your data. Figure 14.9 shows a scatter graph. It may be appropriate to fit a line to the data in order to extract relevant parameters (see *Chapter 13: Numbers, Errors and Statistics*). Figure 14.6 shows an example of a linear fit to a set of data points.

Histograms

These are useful for sets of data that are discrete or binned in some way (for example, where data are grouped by the day of the week). When drawing histograms, state clearly what you are plotting and the units and scale, and make the columns of equal width – this looks better, and for certain calculations you may wish to take into account the area covered by each column (see Figure 14.10). The bars in a histogram should not touch unless they represent continuous data – for example, Figure 14.10(b) could be presented with the bars touching, but not Figure 14.10(a).

Do not spend ages creating amazingly intricate 'three-dimensional' graphs if you have simple data to present (see Figure 14.10(c)). The more complex a graph is, the more difficult it is for the reader to interpret, and the more you will irritate your examiners (we speak from personal experience of being irritated examiners).

If you are comparing normal or control or 'wildtype' (to use a genetics term) data with other data then always plot the normal data first on the histogram, so that it presents the basis from which to then compare your other data (see Figure 14.11).

Pie charts

Pie charts are a simple way of showing proportions and percentages (see Figure 14.12). Again they need to be fully annotated so it is clear what each 'slice' of the pie represents and what proportion or percentage of the whole it represents. Sometimes people like to shade the individual sections of a pie differently, but this can be messy if it is divided into many parts.

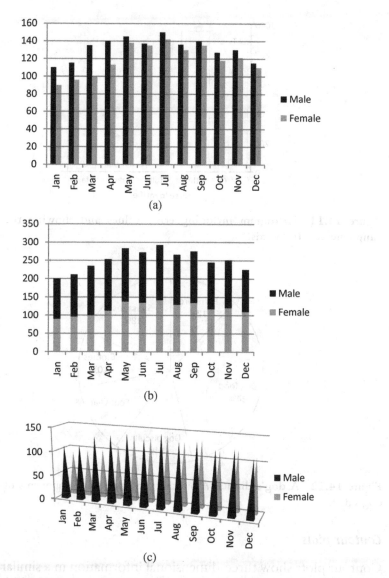

Figure 14.10 Histogram of the number of landfish observed throughout the year. (a) The data are presented as a simple histogram; (b) the same data, but this time the data are stacked; (c) the same data but with a 3D effect, which is visually confusing (and will irritate your examiners).

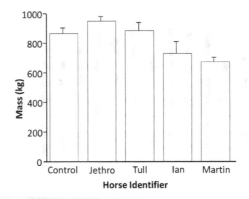

Figure 14.11 Histogram including error values and showing the control sample nearest the *y*-axis.

Figure 14.12 A dog's dinner. A pie chart showing the components of a typical dog's diet.

Contour plots

Contour plots show three-dimensional information in a similar way to a map, by giving contours of equal height (the *z*-value) as a function of the *x*- and *y*-values. A simple example is given in Figure 14.13. Many different varieties, including different viewpoints and various false-colour scales, are possible. Remember always to include a key, and don't let yourself get carried away with all the fancy effects that are available.

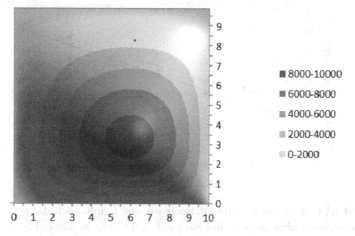

Figure 14.13 A simple contour plot.

Working with multiple datasets on one graph

As a general rule it is a good idea to work with one dataset per graph. Sometimes, however, especially if you are comparing data, it makes sense to plot multiple datasets on one graph, in which case use different symbols for each set so that they can easily be distinguished – dots for one set, crosses for another set, triangles for another, or, perhaps, use different colours if you can. To avoid confusion do not use very similar symbols in the same graph: for example, using small circles and octagons for different datasets is not a good idea. Label each line, but if this would make the graph messy, give a key as to what the symbols represent, as above, or explain them in the figure legend (see Figure 14.14).

Sometimes it is better to produce more than one graph, rather than to superimpose many datasets. If you are showing similar data on more than one graph, try to keep to the same scale so the reader can easily compare them. For example, if you have room on one page, you can keep your *x*-axis constant, but you can lay however many *y*-axes you need against it, clearly separated, but in one column. Beware of making a graph so complicated that it is undecipherable.

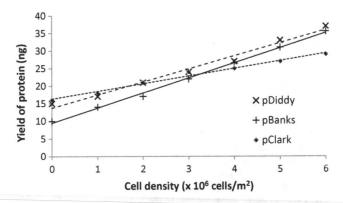

Figure 14.14   Plotting multiple datasets on one graph. Each dataset is plotted with a different symbol and each line has a different format so they can be identified.

14.5 Tables

Tables allow you to present a large amount of data so that various measurements or statistics can be compared. Unlike graphs, in a table you show the exact values of your data. You need to make your tables easy to read. Decide what you want the table to say, and then decide on the best way to say it.

Some basic guidelines

Creating tables

For simple tables with few entries, it will probably be easiest to use your tab key to create the columns. If you are unsure about the use of tab keys see *Chapter 2: Getting Organised.* The tab command will let you align your text to the left, right or centre. Normally tabs align to the left, for example:

Mouse	Cheese	Bacterium	Milk
Cat	Mouse	Cheese	Bacterium
Dog	Cat	Mouse	Cheese
Cow	Dog	Cat	Mouse

This may be useful for certain tables, for example, where the entries are words. However, if you are using your table to present numbers, it is best to align to the right, as this makes it easier to compare and process them.

1998	716	298	1678
14	9	19	59
25	10	1960	1
304	109	1	782

If you have numbers with decimal points, the point should be positioned in the same place in each column. If you have values of less than 1, always put a 0 in front of the decimal point so that your figures read '0.9', '0.043' etc., not '.9', '.043'. Your word processor will format numbers automatically this way if you set it appropriately.

1998.01	71.60	2.98	16.78
1.42	9.00	19.00	0.59
0.25	0.10	19.60	0.01
30.45	110.96	1.67	78.29

For more complex tables it is easier to use the 'Tables' function in your word processor, or to use a spreadsheet program, which will allow you to set up and edit tables quickly and easily (see *Chapter 2: Getting Organised*). They can then be imported into the main document.

   If you have a zero reading, enter it as 0. Different disciplines have different conventions for what to write if you have no reading – most people either leave a blank space, or put in a code such as 'ND', which should be explained in the legend ('ND: not determined'). Do not use a dash '-' as this may be confused with a minus sign '–'.

Table layout

Think carefully about the order of columns and rows; place any data that can be compared in adjacent positions. If you have related numbers, put these in columns so that people can easily scan up and down, rather than from side to side (see Figures 14.15 and 14.16). Arrange columns and rows of related data together in your table, in a logical order. Don't include any more information than is necessary as it will make it harder to take in the information you do want the reader to understand. Try to avoid having a page break in the middle of a table if you can; if this is unavoidable, repeat the column headers on the second page.

Enter your information in an appropriate form using correct scientific notation. Clearly state what units, if any, you are using and use SI units and standard abbreviations. Add in error statistics for numerical values if necessary. As with other places where you present results, do not use too many significant figures.

Annotating tables

Tables need annotating, but keep the annotation as simple as possible so that your tables do not look cluttered. Each column and row should be arranged in a logical order and labelled, explaining its contents and giving units of any values you are entering. If you have long headings, use two or three lines to fit them into the space above the column, or beside the row. It is not necessary to draw lines between columns and rows: doing so often makes the table look fussy. There is no reason why every column should be as wide as every other column, or every row should be as deep as every other row. Space the rows and columns according to the amount of text they contain (see Figure 14.15).

Name	Colour	Age (years)	Ear length (cm)	Number of dog biscuits eaten
Bonzo	Black/brown	2.6	10	45
Rover	Brown	5.7	15	78
Felix	White	5.9	6	2
Henry	Brindle	8.4	12	83
Toby	Black/white	9.1	20	169
Groovy	Orange	10.9	17	50
Tim	Black	15.2	14	39

Table 4.5   Details of the seven dogs taking part in the study.

Figure 14.15   A well-laid-out table. The columns are not equally spaced apart, but look pleasing to the eye, and the numbers are justified to the right and so are easy to read.

Food	Tail wag (per minute)	Bark (per minute)	Jump (per minute)
Dog biscuits	120	83	12
Dog food	102	52	8
Doors	68	34	7
Footwear	116	72	9
Grass	1	1	0
Jumpers	9	5	2
Sausages	150	119	15

Table 5.2   Food stimulus response in dogs.

Figure 14.16   Another well-laid-out table. Careful thought has been given to the column sizes and titles.

Numbers and titles for tables

Each table should have a number and a title (and possibly a legend) which clearly and concisely indicates the contents. Keep it as simple as possible (see Figures 14.15 and 14.16). The table title is listed in the List of Tables at the front of your manuscript (see *Chapter 10: The Other Bits*).

   As you will have noticed, we have carefully organised our tables so that in Figure 14.15 the entries are by dog age, and in Figure 14.16 the entries are in alphabetical order. Tables need logical organisation so that the reader can make sense of them – so have a think about what's the most sensible way to present your data, i.e., what are the important points you wish to emphasise in your table. Always put data into tables in alphabetical and numerical order, if you can.

Footnotes for tables

If you need to add footnotes to your table it is best not to use numbers as symbols because these may be confused with your data. Use asterisks and other symbols: *, §, #, +. It is best to put them in as a superscript so they stand out.

14.6 Inserting Figures and Tables into a Thesis or Dissertation

Cite each figure or table in the text; do not just shove it in and hope the reader realises what it is doing there. Check you have referred to each one in the text, and that there is a figure or table present for each one that is referred to.

Insert each figure into an appropriate place in your text. If possible put your figures and tables in 'upright' so the reader will not have to turn your thesis or dissertation on its side to read them. If this is not possible put them in so the bottom of the figure is always towards the right-hand side of the text. If you try to put the bottom of the figure on the left-hand side, the figure legend will probably be lost in the binding (see Figure 14.17).

Most word processing programs allow you to import figures and tables easily from other types of programs, such as graphics programs or spreadsheets. This is the best way of incorporating figures and complicated tables into your text. Check carefully, on a printout, that no glitches occurred when the figure was imported.

Where to put figures and tables

When you are reading text that refers to a particular figure or table, it is irritating if you have to keep flipping backwards and forwards between pages to look at it. Try to keep your figures and tables as close as possible to where they are discussed in the text. Do not put all your figures and tables at the end of each chapter, because it means the reader is continually having to flip backwards and forwards between the text and the illustrations and it is ... very irritating ...

Including additional figures and tables on a CD or DVD

Sometimes data need to be presented that cannot be shown on paper, for example, audio or video files, or very large bodies of data that need to be submitted on a CD or DVD. If you have to include a small flat object such as a CD or DVD, you can make – or ask the binder

Genomic DNA

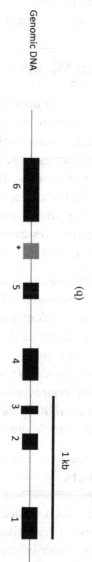

Figure 6.3. Diagram showing the intron-exon structure of the *Mylo* gene. The possible new internal exon is shown in grey with an asterisk * (adapted from MacInnes, M., 2016).

(a)

Figure 14.17   Two figures presented sideways on. (a) is easier to read than (b).

Genomic DNA

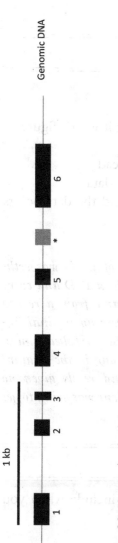

(b)

Figure 6.3. Diagram showing the intron-exon structure of the *Mylo* gene. The possible new internal exon is shown in grey with an asterisk * (adapted from MacInnes, M., 2016).

to make – a pocket at the back of the thesis. Make sure that any objects you submit as part of your thesis are clearly labelled. The labelling must include your name, your department, the date and which degree you are studying for.

Common Mistakes

- Leaving the figures till the last minute
- Not referring to a figure or table in the text
- Referring to data in the text but not illustrating it with a figure
- Omitting scales or scale bars on figures
- Lettering figures with text that is too small to read
- Using cluttered graphs or tables with too much data
- Presenting graphs with lots of clear space beyond the data points because the axes extend too far
- Spelling mistakes in the figure legends.

While writing this book a colleague came to see one of us – he was seething with annoyance: he was examining a thesis for a PhD and every figure the student used in their Introduction came from a review article and although properly referenced, there were no original figures at all from the student; further, some of these published figures from one particular article were quite clearly wrong in the original published review, indicating (a) the student had really given no thought at all as to their contents and (b) the student was going to get a roasting at the oral examination ...

Key Points

- Decide which figures and tables you need to include when you write your thesis plan
- Prepare a draft of each figure and table
- Annotate figures and write a short title and legend
- Label each axis of a graph and add units and scale

- Make sure that all labels, scales, legends and annotations are readable
- Always present your 'normal' or 'control' data first on a histogram or in a complex figure
- With multiple figures for the same samples, present the samples in the same order and standardise as much as possible throughout the thesis or dissertation
- Make all your figures and tables as simple and clear as possible
- Use colour thoughtfully.

Chapter 15

Proofreading, Printing, Binding and Submission

You have already put a lot of hard work into writing your dissertation or thesis. The final steps, printing, binding and submission, are fairly straightforward but still require close attention to detail. It is a shame to spoil an otherwise good text with mistakes in spelling, and layout that is badly thought out.

15.1 Getting Ready

Know your submission date

If you are an undergraduate or master's student you will almost certainly have a deadline for submission set by your department or university: make sure you know this date! It is surprising how often people do not know the timetable of their course.

PhD students do not normally have a set deadline for submission, but some universities have a date by which you must have submitted your thesis in order to be awarded the degree – check with your local administrator for further information (see Figure 15.1).

Whether you set your own deadline or have one set for you, draw up a realistic timetable including enough time for printing, proofreading and binding. Binding alone can easily take a couple of days, possibly even longer for a PhD thesis.

Figure 15.1   Know your submission date (and the date of your viva).

Know how many copies you need

For most undergraduate theses or dissertations you will be required to submit only one or two copies; for PhD theses up to six copies may be needed. Alternatively, you may just need to send a PDF by email. Know what you need.

Figures and tables

It is amazing just how long it can take to add your figures and tables to your text, and just how muddled the process can become when you are tired. To streamline the final production of your thesis, prepare your figures well in advance. Make sure you have any additional material you will be adding by hand prepared and ready for insertion (see *Chapter 14: Figures and Tables*).

Advance organisation

Most people who have written a thesis or dissertation have experienced the frenzied last-minute rush for a deadline. Generations of supervisors have said to generations of students 'Leave yourself enough time at the end'. Generations of students have ignored or forgotten the advice and had to spend the last week or so before their submission date sleeping at their desk, if at all, as they work through the night to get their theses finished. People find themselves printing the final version of the dissertation or thesis at 4 am, only to discover that the printer runs out of toner, computers and scanners mysteriously stop working, and the entire departmental stock of paper runs out because all the other students are printing their theses at the same time. To avoid this nightmare, prepare in advance.

Long ago, when word processors were only able to undo the single most recent command, the spouse of one of us was preparing a dissertation. One of the final steps was to order the references alphabetically. Unfortunately, in the last-minute rush at about 2 am it ended up with the whole dissertation being ordered alphabetically, word by word! *One wrong move and the entire dissertation would have been lost. We still go cold at the thought of what might have happened. The moral is: don't do things in a rush at the last minute. And, of course, keep multiple copies of all important documents.*

15.2 Proofreading

Once your text and figures are ready and your thesis or dissertation is laid out properly, allow yourself enough time for some final checks. Spell-check everything – including figure legends and titles, and the annotations to figures and tables. Spelling mistakes will give a bad impression, and you may have to go back and make whatever spelling corrections your examiners request, before you can be awarded your degree.

Check for contractions like 'it's' and 'isn't' and remove them – these abbreviations are not used in formal scientific writing (try doing a 'Find' command for apostrophes in your text). Also keep an eye out for commonly confused words such as 'proceed' and 'precede' (see *Appendix 1: Easily Confused Words*).

Remember to check the References are complete and have all the relevant information and consistent abbreviations. This can be a painful and time-consuming task, but any errors will be easily spotted by your examiners.

Before you finally print your text, make sure it is properly laid out and looks beautiful by using the 'Print preview' facility of your word processor. This lets you see what each page looks like on your computer screen before committing it to paper. Check that the layout of the thesis is consistent – that your titles are all in the same font, and similarly, of course, *your text is all in the same font.*

Also check that all the heading, table and figure numbers are correct and that all your titles have text underneath them. Something that many people miss before printing is having a title on the last line of a page and the text starting on the next page, which looks awful. Check that the page numbers have come out correctly and that the figures and tables have appeared in the right place in the text.

Once you have looked at your text on computer, print one copy of your thesis. Check that everything looks consistent and is well spaced and lovely, and that you do not have blank pages or other unexpected mistakes. Now thoroughly proofread your text.

Proofreading your text means checking it for mistakes (from grammatical blunders to minor typographical errors), before submitting it. Do not rely on your spell-checker; it will only pick out combinations of letters it does not recognise as words. It will not pick up grammatical mistakes, words that have been used incorrectly or words that have been missed out. Make sure all your numbers are correct; it is very easy to put a decimal point in the wrong place. If you are using equations be absolutely sure they are right. At this stage do not make major changes to your text – stick with what you have written.

It is possible to proofread the text yourself but much better to have someone else proofread it as well, for instance a relative or a friend. It doesn't matter too much if they are not a scientist as many of the grammatical and spelling mistakes that you may have made will be clear even to a non-specialist. They will definitely find mistakes that you have missed. Choose someone who has a good command of English. Once your text has been proofread, make your final corrections. Do not rush this stage just because you are almost finished.

Also at this stage make a final check of the References: in particular, make sure that there are no citations in the text to references that have now disappeared from the Bibliography.

15.3 Printing Your Dissertation or Thesis

For some universities, it is now only necessary to submit your thesis electronically as one PDF file, so you can skip the printing stage for the final version. This has the advantage that you don't have to worry about the details of binding. However, for many people it's still necessary to engage in this final battle of wills with computer technology ...

First, prepare adequate stocks of paper, photographic paper, toner and any other supplies you may need (including strong coffee), so that the final stages in the production of your thesis go as smoothly as possible. It is also worth finding out where alternative computers and printers are, just in case you need them.

Before you print, clear away any old drafts of your thesis that may confuse you and any other debris: crusty old coffee mugs, the remains of your last three meals – anything that might spoil your handiwork. You may be tired and accident-prone and it is very easy to knock a cup of day-old coffee over your final draft or drop your beautifully prepared photographs into a bowl of cornflakes just as the last page of your text is crawling out of the printer (see Figure 15.2).

Check, again, that you have a stack of clean paper in the printer and the toner is not run down. You can now give the 'Print' command to print the final version of your thesis. Print one copy and do a last-minute check for any mistakes before printing further copies.

You will almost certainly have to print more than one copy. Most printers will collate multiple copies for you, but check how your print command works so you do not have to spend hours collating by hand. Make sure that each copy of your thesis or dissertation is complete and in the right order. Students have been known to submit two Chapter Ones and no Chapter Two. Insert any additional sheets that you have prepared such as glossy photographs.

Figure 15.2 Clear away the debris before you print the final version of your thesis or dissertation.

Now place each copy of the thesis in a folder and, if necessary, go to bed.

Including your own publications

Copies of any of your own papers related to your research project can be bound into the back of your thesis or dissertation. You can include papers that are 'In press' (accepted for publication, but not yet published) or that are 'Submitted' (currently being reviewed for publication), providing that you clearly state 'In press' or 'Submitted' on the Title Page of the paper.

Binding

You now have a very valuable stack of papers. These need to be bound together so that none are lost and your thesis or dissertation is easy to read and handle. How your text should be bound will depend on your department and your degree.

You must check with your department or university about rules covering binding. For an undergraduate dissertation, for example, you could buy a plastic spine that slides along the left-hand side of your stack of papers to hold them in place. A sheet of transparent plastic, such as the type used for overhead projectors, at the front and back of the text will make it look neater and help protect it. Some departments have access to binding equipment in their administrative offices which produce either spiral bound or glued theses. If such a service is available to you, find out in advance who runs it and make sure that they are ready for your thesis, especially if you are going to be in a rush to get it submitted. This type of binding is usually fairly speedy, but if an entire class is waiting, you do not want to be last in the queue.

For higher degrees such as MScs and PhDs you are usually required to submit a professionally bound text. Some institutions will allow you to submit a text in a temporary binding; however, the text will have to be properly bound once these corrections have been made. If you have to submit a professionally bound text find out the names of recommended local binders. Your university or department should be able to provide you with a list. Professional binding can be expensive, but often binders will reduce the charges if you are not in a hurry and can wait for them to fit in your job when they have little else to do. Telephone the binders to check what they charge and where they are.

For the higher degrees, universities usually have strict rules about the binding of the thesis and its lettering: what colour the cover should be, the font and colouring of the letters and what it should say – for example name, degree title, date. Many experienced professional thesis binders will know the rules for their local university or college, but you should make sure you know the rules yourself. Ask your supervisor or departmental administrator.

15.4 Submitting Your Thesis

Submitting your thesis is the easy bit. All you have to do is give it to the relevant authority for examination. The only things you do have to be absolutely sure about are who and where the relevant authority is. The Finance Department will probably find your text very useful as a doorstop, but they are not authorised to award you with any academic qualification. Find out from your supervisor or administrator where you need to go and what their office hours are so that as soon as your thesis is bound you can hand it in.

Even if you have to submit hard copies of your thesis, you may need to submit an electronic copy as well so check again if this is necessary.

What happens next

What happens next depends on the degree for which you are studying. For undergraduate and some master's courses, your dissertation or thesis grade might simply be part of your final mark. For doctoral students you may wait several weeks for an oral exam, a viva, in which your thesis is scrutinised by two experts (see *Chapter 16: The Viva and Thereafter*).

Common Mistakes

- Not allowing enough time for the final stages of preparing a thesis or dissertation
- Not thoroughly checking the final version for errors.

Several years ago, a friend, who is now a professor at a major Californian university, was running very close to the submission date for his BA dissertation and it looked as if he would miss this deadline. He had arranged with the friendly college librarian for her to bind the thesis for him. By midday of the last possible day he could submit the work he was still printing it and sticking in his figures. By 4.30 pm he had finished. He collected all his papers together, ran from the computer room to the library, up the steps and into the arms of the waiting librarian. She grabbed all four copies of the thesis, shoved them through the binding machine and handed the bound copies to our flushed and nervous hero. By this time it was 4.45 pm – the deadline was 5 pm. He then ran from the library to the submissions office, arriving at 4.55 pm. He had five minutes to spare and nonchalantly handed the four copies of his text to the administrator, who opened the theses to find all four copies had been bound down the right-hand side of the page.

Key Points

- Find out about submission rules and dates
- Do not underestimate the length of time required for the printing and binding process
- Spell-check, proofread and check the layout of a printout of your thesis or dissertation before printing the final copies
- Carry out one final check of copy of your manuscript before it is bound.

It is always worth checking that the paperwork – in this case your thesis – has been received and all is well. One of the authors of this book was sent some finals exam papers to mark, by recorded delivery, from the other author's university. The exams were despatched and arrived on time. And were then placed under the shelf in the post room of the receiving institute for a … very … very … long time … before anyone realised they were missing.

Chapter 16

The Viva and Thereafter or 'And you Thought it was All Over ...'

16.1 Oral Examinations

There are two reasons you might have to take an oral examination, commonly known as a viva (an abbreviation of the Latin *viva voce* meaning by, or with, the living voice). At undergraduate level (bachelor's) and master's level in the UK, you might have to take a viva if your examination results are borderline, for example, between pass and fail, or first and second class. Alternatively, if you are writing a dissertation or thesis, a viva may be part of the examination process that everyone has to go through – depending on the institution and degree. In this chapter we are only considering those vivas that concern theses or dissertations and are a set part of a degree course. This will apply to nearly all PhD courses and some MSc courses. We are looking at a typical viva in which the student is in an oral examination with two examiners, but many universities have different systems, some involving a public defence of the student's work – in this case, similar preparation to that described below is useful.

As with any other examination, you are more likely to pass if you prepare properly for the viva and hone your technique. The examiners are not there to make you fail. They will probably try to make the viva as enjoyable and interesting as possible both for you and for themselves. Some people do actually enjoy their vivas – for a PhD student who has poured their heart and soul into a project for three years or more, the viva is probably the last chance to discuss their work with experts who have studied it in detail.

Your examiners will be trying to make sure that you have really thought about your project, its data and implications, and that you have an understanding of your field appropriate to the level of your degree. Just as you are more likely to convince your examiners of your professionalism if your thesis is well planned and well presented, you are more likely to convince them of your worthiness at the viva if you plan for it and present yourself well.

What to expect in the viva

All vivas have a similar format, whichever degree they are for. Your university or department will arrange a venue and the time. You enter the room and are questioned about the contents of your thesis and background knowledge. For a short BSc thesis this might last ten minutes; for a PhD thesis it could take a few hours. The examiners will tell you when the viva is over. It is a gruelling experience for both you and your examiners.

Often a viva will start with a general question asking you to summarise your work or to discuss the highlights of your results. It's a good idea to practise giving a two-minute summary of what you have done so that you can answer a question like this confidently, and efficiently.

For most vivas you will then be asked questions covering your work, your general field of research, and the background to your project. Questions about your project will be both theoretical (why you took certain approaches, why you developed certain strategies, what the principles behind certain techniques are) and practical (exactly how you carried out your experiments, and the detailed interpretation of your data).

The level of knowledge that you are expected to have will depend on which degree you are taking. PhD students have had at least three years more study than undergraduates and thus will be expected to know their chosen field in greater depth.

16.2 Preparing for Your Viva
Know your thesis

A cliché that is often used to make new seminar speakers feel less nervous is that 'you know your own work better than anyone else'. Like most clichés, it is generally true, and it applies to vivas as well as to seminars.

Read through your thesis carefully several days before your viva. As you are reading, think what questions the examiners might ask. Note these questions on a sheet of paper and make sure you can answer them correctly. This will give you some idea of how much you really know: if you find yourself guessing or floundering on any of the questions, revise this area of your knowledge. Think about why you took certain approaches, and whether you would do the same again in retrospect. Also think about what future work might come from your thesis.

Remember that everything you have written in your thesis is fair game for a question from the examiners. So be prepared for questions on anything you included.

Your viva is an oral examination so it is a good idea to practise answering the questions out loud. Your flatmates will have got used to your strange behaviour by now, so they will not think it unusual to hear you talking to yourself. If you can, get a friend to run through the thesis and ask you questions. You can also ask your supervisor to arrange a 'mock' viva, so that you get some feel of what the real thing is like. Not all supervisors will agree as it is a large amount of work for the 'mock' examiners, who simply may not have the spare time, but it is worth asking.

Discuss your project with your supervisor before the viva. Ask them for comments and criticisms and tell them about your ideas and your response to likely questions; your supervisor should be a helpful sounding board.

Make a list of errors

If you have found errors in your thesis or dissertation, either typographical or in the data, make a note of them and bring a list to the viva. The examiners will appreciate that you have thought about your thesis and are honestly trying to produce the best possible piece of work.

Know your field

Your examiners will not confine their questioning to your thesis or dissertation; they will also examine, to some extent, your knowledge of the field in which you work. You can never know exactly what the examiners will ask, but you can feel prepared if you have considered the following questions:

- What research and what ideas led to the creation of your field?
- What research and what ideas led to the creation of your project?
- What questions has your project answered?
- Have similar projects to yours been undertaken before, and what have these told us about your area of research?
- What have similar projects told us about the techniques you used?
- Are there other techniques you might now use?
- What are the implications of your area of study for other fields, for example, how does biotechnology affect environmental sciences; are any ethical issues involved?
- Does your work have any industrial or medical applications?
- Have any key papers been published in your area since you submitted your thesis or dissertation?
- What future work could come out of your project?

Know your References

Remember to read your References. You could legitimately be asked a question about any one of your cited references. Make sure you really do know what is in the key papers you have used. It is remarkable how often people cannot answer a question about a paper they have referenced many times in their thesis; possibly because they are genuinely too nervous to remember, possibly because they took the reference from a website, without reading the paper.

Make sure your knowledge is up-to-date

Murphy's law requires that any major breakthrough that occurs in your field and directly affects your results will happen on the morning of your viva or possibly three or four days before. So make sure that your reading is up-to-date and you have a very good idea of what is happening currently in your chosen research area. You might not feel like it but, particularly in the week preceding your viva, go to the library and read any publications that might be relevant. This is an absolute must if you have had a long break between carrying out the research project and reaching the viva.

Know your examiners

You may know in advance who your examiners are. If you do, it is worth finding out as much as possible about them (see Figure 16.1). Carry out a database search and find out what they have published and what their scientific interests are. This may give you some idea of their likely line of questioning ... but do not be lulled into a false sense of security: it is dangerous to assume too much.

Figure 16.1 Know your examiners.

Formats of undergraduate and master's vivas vary, but most PhD and DPhil vivas are undertaken by two examiners: an 'internal' examiner from your university (but not your immediate research group) and an 'external' examiner from outside your university. The external examiner is usually the person leading the viva. Sometimes the supervisor may be present as an observer but this is unusual.

The night before the viva

Although a little light reading will probably not hurt you, it is too late to cram in much more information the night before your viva. Relax and get yourself in the right frame of mind. Put on a favourite album. Do not go out and get paralytically drunk with your friends. Do not get arrested. Do not have a row with your partner. Do not decide to murder your supervisor – what is done is done. Have a nice meal. Make sure you know exactly where the viva is and how long it will take you to get there. Make sure your clothes for the viva are ready to wear, are comfortable, and make you feel good. Make sure you have enough money to get to the viva, and you have a phone for any calls you want to make after the viva. Make sure any paperwork or data you want to take with you are prepared. Also make sure you have a bottle of water, packet of tissues, throat lozenges, or whatever else you might need in the viva (see *Things to take with you*, below).

Relax. Remember that (1) many people have been in your position before and have survived, and (2) the examiners know perfectly well how you feel, and are generally sympathetic.

16.3 The Day of the Viva

Finally the day of the viva comes. Be sensible and realistic about travel arrangements and don't leave anything to chance.

It is a good idea to arrive about 20 minutes early and to let the examiners know that you are there. Take a few minutes to get your bearings, then head for the nearest lavatory. If you are due for a long viva you want to feel as comfortable as possible. Sitting cross-legged for five hours because you forgot to go to the toilet beforehand is not feeling comfortable. Toilets have mirrors, so you can check you look terrific and all your usual bodily parts are still there. You can also give yourself a big encouraging smile and check that your shoes are on the right way round and that you look respectable. It is difficult trying to answer a question about quantum mechanics when you suddenly notice you are wearing odd socks and have a toothpaste stain on your shirt.

A few minutes before the viva, wait where the person in charge of the exam can see you. You will be called into the examination room (see Figure 16.2). Walk in. Sit down. Keep breathing. The examiners will introduce themselves, and then, almost always, start with some fairly gentle questions to get the ball rolling. There is no need to burst into tears as you enter the room. The examiners know how you feel. Take a deep breath and off you go!

If you feel nervous about your viva remember that almost everyone does – especially those who have to present a public defence of

Figure 16.2 Finally the day of the viva arrives.

their work in front of family, friends, and the rugby team, as many students across the globe have to.

Things to take with you

Take a copy of your thesis or dissertation to the viva so you can refer to it when asked questions. Make sure you know your way around the thesis so you can find things easily. Stick pieces of paper into it so that you can find important bits quickly, if you need to.

Occasionally examiners will request that a candidate bring along some of their raw data to the viva, such as photographs or traces (see Figure 16.3). Even if you are not asked to, you might want to bring in raw data, your practical books or other material that relates to results your examiners may call into question. For example, if important details in a figure are difficult to make out because of problems with the original or with reproduction of it, you might want to bring the original so you can show the examiners that a detail such as a faint image, spectral line or other feature is really present.

Figure 16.3   Bring raw data to your viva if necessary.

In the viva

Sound enthusiastic. We knew a student who was extremely bright and a very good scientist. She did not come across well in interviews or oral examinations because when she was nervous she appeared detached and uninterested. Interviewers felt she was bored by the topic and probably did not know much about it. In fact she had a passion for her subject and has gone on to a very successful academic career. Another student, probably not as able, did extremely well in oral examinations because he sounded enthusiastic about his project and so endeared himself to his examiners (who are only human, after all). You chose your subject because something about it appealed to you. Remember that enthusiasm and pass it on to the examiners. Enthusiasm is infectious. If you seem interested in your work they will be interested in both it and you.

If you are asked a question and have absolutely no idea of the answer, it is best to be honest rather than to try and bluff your way through. Bluffing is usually obvious and may get you into deeper water if the examiner decides to pursue your answer. If you disagree with an examiner, then by all means say so, providing you can justify your point of view.

If you don't understand a question, ask for clarification. You don't want to waste time answering a different question from what the examiner asked.

16.4 After the Viva

For some exams you have to wait several days to find out your results. For others, such as a PhD, you will know the result the same day. For undergraduate degrees, such as a BSc, BEng, BA, MSci etc., finding out your result will be the end of the story.

For PhDs and a few MScs you will almost certainly have corrections to carry out – perhaps simply typographical errors but often including other changes. Your corrections, however trivial, must be checked and passed by an examiner before they can recommend the university to award you a degree.

Do your corrections immediately, because anything left for a long time becomes more of a chore and because universities often have strict rules on how much time is allowed for corrections. For example, most universities in London require minor PhD corrections to be seen and checked by an examiner within three months of the viva – if the student overruns this time limit they may have to have ANOTHER VIVA ... (which will cause you a lot of unnecessary stress and *really* annoy the examiners ...)

If you have to carry out further work such as more experiments or additional analysis of your data, liaise with your examiners about exactly what needs to be done, and about resubmitting your thesis.

Very rarely people feel the result of their viva was inappropriate. If you feel that your result is unfair take up the matter with your supervisor and university or departmental authorities immediately, who will look into it.

If you should fail

In the highly unlikely event that you fail your degree, do not despair. Remember that you always have a variety of options for dealing with any situation, however depressing or embarrassing or even hopeless it may seem. You need some time to think about what you want to do next so don't rush any decisions. Other people have been in your position and have gone on to great success, really, they have.

First find out your options, of which there will be several. Talk to your supervisor or academic administrator to find out what academic routes are open to you. Discuss them with family and friends and other people who know you. University counsellors are used to helping people with decisions and will be able to help you. Do not lose faith in yourself and remember that any problem always has more than one solution. Give yourself some time to look at different possibilities.

Common Mistakes

- Not reading the thesis beforehand
- Not knowing the references that have been cited in the thesis
- Not thinking about the background to a project
- Not planning travel to the viva carefully and not allowing for delays.

One of us had to struggle through heavy snow to catch a (delayed) train for a viva in a distant city. The other examiner flew in from another (snowy) country. But the candidate had ignored advice and gone home to his small village a few miles away in the local hills the night before the viva. He was completely snowed in the next morning and the viva had to be cancelled. This left the two examiners stuck in the city, having wasted considerable time and money to get to the viva. They were not pleased with the candidate ...

Key Points

- Prepare yourself for the viva: read your thesis and know your way around it
- Think about likely questions and ask your supervisor for comments
- Think about the background to your research project
- Revise the key references that you have cited
- Ensure your reading is up to date
- Use your common sense and remember that lots of other people have survived the process.

Do not forget our advice about checking the mirror before you go into a viva. One of us was involved in examining a PhD student who was extremely nervous. He walked into the viva wearing a brilliant white shirt and black trousers – with his flies undone. The examiners were in agonies of indecision about whether or not to tell him. They decided not to as they thought it would make him more nervous than he already was. About 45 minutes into the viva the student noticed his flies were undone and spent the rest of his viva in an embarrassed haze trying to both answer the questions and decide whether or not to discreetly reach for the zip.

Chapter 17

Supervision

Your supervisor is there to criticise your work on your thesis or dissertation constructively. Before starting on your project, particularly if you are an undergraduate, you might only have seen your supervisor from a distance, probably giving a lecture during which you could sit passively and either listen (or not listen) to what they had to say. Your dealings with them as a supervisor are completely different. You are an active partner in a two-way relationship. If you contribute little or nothing, your supervisor will have little or nothing to supervise. The longer and more complicated your project, the more important your relationship with your supervisor is. Like any relationship, both sides have responsibilities and need to communicate effectively with each other. Like any other relationship, there are probably going to be times when you don't get on with each other.

17.1 Roles and Responsibilities

Your role

Your supervisor is a human being and you need to engage with them, and to take responsibility for your role in your project. In this way you 'take ownership' of your project and you will gain confidence, even if the results are not always necessarily what you want. 'Taking ownership' is very important for you and your project.

Your supervisor will be responding to you and your work. If you come to the relationship with an enthusiasm for your work and a willingness to discuss it, your supervisor is more likely to take an interest in you and your project and therefore give better advice and criticism.

You are not expected to work in complete isolation and you should receive support from your supervisor; on the other hand try not to abuse this relationship by pestering supervisors every five minutes with trivial problems that you could solve yourself. It is up to you to order your ideas and grapple with interpreting your experimental or theoretical results.

The role of the supervisor or tutor

Your supervisor should provide helpful comments about how to sort the main ideas and themes of your work and structure your thesis. They should be prepared to read through drafts of chapters and the final version of your thesis or dissertation, and provide constructive criticism about the appropriateness of the contents, the level of understanding you have shown, the thoroughness of your analysis, and the justification of your conclusions. They should also be able to advise you on your use of English and recommend suitable graphics and reference database programs to use. Having said this, your supervisor is not instantly available for you 24 hours a day. They will have their own timetable and deadlines, and you need to give them time to read your writing.

It is not the supervisor's role to write the thesis for you. Your supervisor is there to guide your project, but they are not your personal secretary and proofreader. At the end of the day it is your thesis and you have to be responsible for its contents.

Second supervisors and thesis committees

Some universities have a system where all PhD students are given a second supervisor or a thesis committee. Their role will be different in different institutions. In some cases a second supervisor will be a close colleague of your main supervisor: someone who is also involved to some extent in the research that you are carrying out. They are therefore likely to know about your project. In this case the second supervisor is someone who you can ask for advice on your research whenever it's not possible for you to see your main supervisor. In other cases the second supervisor may have more of a mentoring role. Alternatively, you may have a thesis committee of two or three people who know about your field but are not undertaking work directly related to yours. However, the committee will be made up of experienced and established scientists who can advise you about the progress of your work, and provide support if you need it.

The second supervisor and/or the thesis committee can be a backup in case any problems develop. They are there for you to ask for more general advice, for example if you are having problems getting on with your supervisor, or if you are worried about the overall progress of your research. Especially at the point where you are writing up your work, you may find the need to talk about your thesis to someone who isn't your supervisor, so make full use of these people.

If your main supervisor is relatively junior in your department, it may be that you are allocated a second, more experienced, supervisor automatically so that someone can keep an eye on the progress of the project and of your PhD. The aim here is to give support to a relatively inexperienced supervisor and make sure that you as the student get all the support that you need. It takes time to learn to be a good supervisor and the involvement of a more experienced member of staff can be of great help both to the new supervisor and to their students.

17.2 Reviewing Drafts of Your Thesis or Dissertation

Make regular appointments with your supervisor to go through your work, and make sure you keep them (see Figure 17.1). Ideally, you should expect to meet once a week at this stage, but this will depend on your individual circumstances. As a minimum, try and meet at least once every two weeks, whether you are studying for an undergraduate degree, master's or doctorate. Give your supervisor a copy of the draft you wish to discuss well before you meet so they have time to read it. Apart from the constructive criticism a supervisor can provide, seeing

Figure 17.1 Colin J. Bodkin (DPhil, pending) remembers he has a meeting with his supervisor.

them regularly will help to keep you motivated to produce your thesis or dissertation on time.

Make sure the draft you show your supervisor or tutor is spell-checked, properly formatted, fully referenced and in a font that it is easy to read: use one-and-a-half or double line spacing so comments can be written easily on the draft. From a supervisor's point of view it is intensely irritating to have to spend time correcting distracting mis-spelling and formatting, which could easily have been done by the student. Your supervisor's time is precious. In addition to looking after you they will have other students to supervise, research to carry out, teaching to do, talks to prepare, letters, grants, papers, reports, and books to write, scientific meetings to organise, committees to chair, papers to track down in the library, papers to referee, journals to edit, phones to answer, collaborators to email, finance statements to work through, staff assessments to prepare, computers to mend, office supplies to fetch, works departments to interact with when their office door falls off its hinges, academic boards to sit on, etc., etc., etc. The list is endless. Trust us, it's *endless*. You are just one part of their responsibilities and you need them to focus on the contents of your thesis, not your spelling mistakes.

For the same reason, try not to present your supervisor with multiple drafts of the same chapter over several weeks. It is much better to concentrate on producing a complete draft of a chapter and only expect your supervisor to comment on it once. It is an inefficient use of everyone's time if you give your supervisor a chapter with only minor changes since the time they last saw it. But when you have a complete draft of the whole thesis, your supervisor will probably want to look at it again, in order to see how it all fits together.

17.3 Problems with Supervisors

PhD supervision is one of the most important aspects of an academic's role, so supporting you through the process of writing your thesis should be high on their list of priorities. However, sometimes things can go wrong.

Some students feel, rightly or wrongly, that they and their work are being ignored by their supervisor – '*I sent my thesis draft to my supervisor two months ago and he hasn't looked at it*'. If you feel that you are getting no feedback from your supervisor, arrange a meeting to discuss the problem. See if there is anything either of you can do to ease the situation; possibly you are being over anxious and a talk with your supervisor may put things into perspective. If, having talked to your supervisor, you still feel they are avoiding their responsibilities and not fulfilling their role, and they have made no indication that they will give more time to your work, then there are one or two things you can do to help get around the problem. Try meeting your supervisor and giving them your complete thesis plan, then set out a realistic timetable with them, detailing when you expect to have different chapters written. This will put your supervisor under gentle pressure. The timetable that the two of you have drawn up gives them deadlines by which they should have read each chapter and returned it with comments. At the same time, try to book a set of regular meetings in their diary for discussing your work. If your supervisor has a secretary, put these meetings into the secretary's diary as well – and of course remember that you must turn up to the meetings that you have arranged.

The opposite problem can also arise (much more rarely), when a supervisor is overzealous and tries to take over the writing of the thesis. It should be possible to resolve this sort of problem through discussion and agreement. If something like this happens, remember that your supervisor is just keen for you to produce the best possible thesis – but you may need to remind them that it is indeed your thesis, not theirs.

Most people survive their degree and their supervisor. In a very small number of cases the relationship with the supervisor breaks down. This might be the student's fault, or it might be the supervisor's, or, more likely, both. It is quite reasonable for a supervisor to get irritable if you mess them around with missed appointments and a sloppy attitude. On the other hand, you may feel they're missing appointments and have a sloppy attitude ... In either case, the first thing to try doing is to set aside time to talk with your supervisor. They may not know how you feel and need to be told there is a problem before they can do anything. Most supervisors want to do a good job but might, for personal reasons, find it difficult to do so. Remember that supervisors are as human as students. They make mistakes. They are emotional. They get divorced, and worry about the meaning of life.

You may feel loyal and sympathetic towards your supervisor, but this will not help you get your degree. If talking with your supervisor does not solve the problem you will have to go elsewhere for help. You could, by agreement with your supervisor, ask another member of the academic staff in your department to supervise you on an informal basis. If this is not possible you could see your department's director of studies (or graduate tutor), who should be able to sort out the problem either formally or informally. Most departments will also have a member of academic staff to whom students can talk confidentially about this sort of problem or who can direct them to specialist university officers. Sometimes students and supervisors suffer from serious personality clashes. By the time students are writing their thesis or dissertation this sort of problem has usually been resolved; if it has not, the student should contact their director of studies.

Obviously, if you are or have been suffering from sexual or racial harassment, or bullying of any sort, then you really do need to let someone know. If you feel uneasy about going through formal channels then talk to a sympathetic member of staff; this could an academic or it might be a counsellor, whoever you feel most comfortable with. Never, ever, simply let things carry on in order to have a quiet life, because generally these problems do not just go away.

17.4 Changing Supervisor

Occasionally students may have to formally change supervisor while writing their thesis or dissertation, for example, if their supervisor leaves the department suddenly. This is usually only an issue for PhD students who may take a few months to write up. It is the department's responsibility to allocate another supervisor. If a supervisor is leaving for another job, this has probably been on the cards for some time and the department should have already lined up a replacement. If the supervisor is suddenly taken ill or dies, the student might be left for a short time without guidance, but a new supervisor will be found as quickly as possible.

Common Mistakes

- Not taking full ownership of the process of preparing your thesis or dissertation
- Making unreasonable demands on your supervisor at the writing up stage

- Giving your supervisor poorly prepared drafts of chapters
- Not seeking help when things start to go wrong.

Key Points

- Remember how little time your supervisor has and streamline the process of reading your thesis drafts – aim to give them a complete draft of each chapter once for comment
- Spell-check and format your drafts and print them in one-and-a-half or double line spacing for easy reading; the supervisor needs to be able to concentrate on what you have written without being distracted by typographical errors
- Remember that ultimately your thesis is your responsibility, not your supervisor's
- Be realistic about the amount of time it will take to complete your thesis
- If you feel your supervision is inadequate then first try talking to your supervisor. If this does not make a difference then you must get help from elsewhere: talk to a counsellor or your director of studies.

A flatmate of one of the authors worked in a very successful and dynamic lab in Boston. The lab head had an unusual approach to management and care of his postdocs and graduate students. On one notable occasion when this flatmate had finished an arduous and demanding experiment and it had not worked, the lab head rolled up a piece of X-ray film and walked round the lab trumpeting 'failure, failure'. If this happens to you, it's probably not unreasonable to feel aggrieved.

Chapter 18

The Use of English in Scientific Writing

To convey your meaning to the reader you need a good command of English. Bad English is not only irritating to read, but can also affect the meaning of what you write. In this chapter we start by looking at the scientific style of writing and the conventions that are usually adopted. We then go on to consider grammar, vocabulary and punctuation. If you are seriously interested in, or worried about, your English, buy a good grammar book, a book on English usage, and a dictionary.

18.1 Scientific Writing Style

One of the most important skills a scientist needs to learn is how to communicate effectively and efficiently. Scientific writing style aims to be clear and precise. When writing scientific English, keep your prose impersonal because the reader is not interested in you *per se*, but in your research. Avoid emotive adjectives because if the research is *wonderful, inspiring, relevant,* or *important,* you do not need to tell the reader – it will be self-evident from the text.

Write in a straightforward style that presents only what you need to say. Keep the words to a minimum and remove any that are not necessary, but make sure the text is easy to read and flows smoothly. If you get stuck, try pretending that you are telling a friend what you have done; use a normal conversational style and then change this into formal scientific writing by removing the colloquialisms and slang and by being completely factual.

Bear in mind that your dissertation or thesis is a serious scientific document, but do not try to force your writing into a style that feels uncomfortable or unnatural. Do not, under any circumstances, be tempted into using long impressive-sounding words in an effort to make the writing sound more scientific. Examiners recognise such ruses, and quantity is no substitute for quality. Abstruse English is boring, pretentious, and dull, and if the words are used catachrestically (improperly) it looks silly as well. Avoid slang and other colloquialisms and, apart from in your Acknowledgements, there is no room for metaphor, irony or jokes.

Remember never to copy anything from other people's work, such as papers, without fully acknowledging and referencing it. Students are failed for plagiarism (see *Chapter 11: Other People's Work*).

Below are some general points to consider about style when writing your thesis or dissertation.

Planning your writing

Just as you should plan the grand sweep of your argument, it is a good idea to plan how to present your information, from paragraph to paragraph, and from sentence to sentence. Make sure your ideas flow rationally and understandably from one sentence to another and from one paragraph to another.

When planning how to write a chapter or section, list the topics and arrange them into a rational order. Once you have the topics in order, you have the backbone of your piece in place. You can now go through this outline noting down the points you wish to discuss under each topic. These will form the core of your sentences. Make sure that your sentences build rationally to a main point at the end of each paragraph.

Headings and titles

Choose short self-explanatory headings and titles to separate the different sections of your thesis or dissertation. Use them to bring out the structure of the document. Don't let sections get too long: use subsections and headings to break them up if necessary.

Planning your paragraphs

The structure of your paragraphs should mirror that of your whole text. They should ideally have a beginning, a middle and an end. Restrict each paragraph to one main idea, which you give to the reader in your 'topic sentence'. The topic sentence will usually come at the beginning of the paragraph and tells the reader what the paragraph will be about. Keep this topic sentence in mind as you write the rest of your paragraph to ensure that the paragraph is coherent within itself. Each sentence should flow logically on from the previous sentence to build your argument towards the end point of your paragraph. If you find you need to include ideas which are not covered by your opening topic, either widen your topic to encompass them, or put these ideas into another paragraph.

Which voice and person to use

One of the first things you will have to decide is which voice (active or passive) and person (first, second or third) to use. Many people think (wrongly) that they have to use the third person passive to sound scientific:

It was found ...

rather than the first person active:

I/We found ...

The passive is used when we do not know or do not need to know who or what carried out the action, which is the case with a lot of science; however, the passive can become very wordy and pompous if used to excess, and there is no reason to conceal the fact that it was actually you who did your research. Our advice is to use the active as much as possible. Compare the following:

Third person passive:	*The digestions were continued for a further five hours.*
First person active:	*I continued the digestions for a further five hours.*
Third person active:	*Digestion continued for a further five hours.*

The third person active is shorter, more direct and easier to read, while the third person passive is quite acceptable when writing about what you did:

> The blue mice were placed in the maze.

Try to avoid the passive voice when you are talking about what other people or things did. Compare the following two sentences:

> The maze was successfully traversed by the blue mice. (*passive*)

> The blue mice successfully traversed the maze. (*active*)

Again, the active sentence does the job just as well and more efficiently.

Which tense to use

You will use a range of tenses depending on what you are writing about. It is best to use past tenses, on the whole, when talking about what you did, but beware of confusing the use of the present simple and the past simple because this will affect the meaning of your sentence. Amongst other things the present simple is used to express habitual actions, 'Tim cleans his teeth every morning', universal truths, 'Water boils at 100 °C', and universal opinions, '£100 is better than a slap in the face with a wet fish'. The past simple is used to say what happened in the past: use it for things that you did, 'I put the chemicals in the refrigerator', observations you made 'A shiny deposit of an unknown material appeared on my bread', and specific rather than general conclusions you came to, 'I concluded that I should not eat the bread'.

Sometimes it does not matter which tense you use, for example, in your Materials and Methods you could write either '5 ml water is added' or '5 ml water was added'. In the first sentence you would mean, 'in the procedure 5 ml of water is always added'. In the second sentence you would mean, 'I added 5 ml of water when I did it'. If you are writing about the result of your experiment, take care: for example, '5 g gold is precipitated' means that it always is, '5 g gold was precipitated' means that it was in your experiment.

Directly addressing the reader

Your thesis or dissertation is a formal document, so avoid phrases like 'You can see the progression in Figure 3.6'. If you have to address the reader, you could use either *we*, 'We can see the progression in Figure 3.6', or if you feel the need to be more formal, *one*, 'One can see the progression in Figure 3.6'. However, it is far better, in a formal document, to avoid addressing the reader directly and to use the third person passive, 'The progression can be seen in Figure 3.6'.

Make sure your writing flows

You might find that the sentences in your paragraph, although all connected to the topic sentence, seem to jump about rather jerkily from idea to idea. You can use transitional words and phrases like *therefore, in particular, but, although, since, because,* etc. to lubricate the changes and show the reader where your argument is going.

If you find that, while they work well in themselves, your paragraphs do not seem to fit very well together, a useful trick is simply to repeat a word or idea from the conclusion of the previous paragraph as an introduction to the next one to show the transition, for example:

> When the pistons move up and down, the connecting rods make the offset crank pins rotate, thus causing the crankshaft itself to rotate.

We see therefore that crankshafts are an essential component in the piston engine.

Another interesting component of the piston engine is the piston itself ...

Comprehensibility

Keep your writing simple, direct and explicit. The following sentence, from a BSc thesis that one of the authors of this book was marking, is very hard to understand:

> The pouring of the gel was immediate as the tendency to set was great.

This probably means:

> The gel was poured immediately because it sets quickly.

Jargon

Depending on your point of view, jargon is meaningless gibberish, or the specialised language of a particular field. Use specialised terms that will be understood by your readers, but make sure you use them correctly, and that they are widely known and accepted. If you have any doubt, give a definition to be safe.

Wordiness

Some words and phrases always tend towards wordiness. Alarm bells should start ringing if you catch yourself writing things like:

> The utilisation of gamma super grass led to improvements in the symptoms experienced by the patients.

It would be more efficient to write:

Gamma super grass improved the patients' symptoms.

It is best, on the whole, to use plain rather than fancy words, for example, *start*, or *begin*, rather than *commence*, *attempt* or *try* rather than *endeavour* or *essay*.

In *Appendix 3: Wordy Words and Phrases*, we have listed words that can cause wordiness, but here are a few of the more common ones with their less wordy alternatives:

... boiled on a heating block **so as** to denature the protein.
... boiled on a heating block to denature the protein.

Washing **of** filters
Filter washing

... the columns were left **sitting** for 2 minutes.
... the columns were left for 2 minutes.

... **the process of** sample preparation involves ...
... sample preparation involves ...

That can also lead to wordiness. It is used more often than is necessary, for example, in the sentence:

There are three things that this depends on ...

that is not necessary. You could just write:

There are three things this depends on ...

However, there are occasions where inserting 'that' can help to make the structure of a long sentence clear, or even avoid possible misunderstandings.

Keep an eye out for tautologies – using more than one word to express the same idea in the same sentence or phrase (for example, *attach ... together, a pair of twins, seems that ... could be possible, repeat ... again, basic essentials*). The following sentence contains a tautology:

The reason my experiment failed is because I forgot to turn on the apparatus.

'The reason' and 'because' are both doing the same job; you could write simply:

My experiment failed because I forgot to turn on the apparatus.

You need to be precise in your choice of words. If there is an appropriate scientific term, use it.

Use distinctions and definitions where they are needed to avoid confusion, but try not to use them unnecessarily. We have bracketed the unnecessary distinction and definition in the passage below:

The Yellow Landfish is a British sub-species of the Continental Blue-headed Landfish. It is a summer visitor (and migratory bird of passage), mainly found in pastures (rather than woodlands, marshes, or mountains).

Vagueness

Your examiners need to know exactly what you did and what you think. Chatting in the lab or over breakfast you might say things like 'The reaction was very quick'. If someone needs to know exactly how quick it was they can ask you. In your dissertation or thesis 'The reaction was very quick' is irritatingly vague. *Quick* is a subjective idea, as is *very*. Is *quick* a nanosecond, or a few hours? Equally irritating are phrases like:

... *less* protein was used.

How much is less protein?

Having dried *sufficiently* ...

What does this mean?

An *appropriate* volume.

What is an appropriate volume? Also be careful about making imprecise comparisons, for example:

> The mauve mouse is *heavier* than the blue mouse.

How much is heavier?

Subjective adjectives and adjectival phrases like these should be banished from your text. It is a good idea to look out for them while you are proofreading your thesis or dissertation. We have included a list of words and phrases to avoid in *Appendix 3: Wordy Words and Phrases* and *Appendix 4: Words That Can Cause Vagueness*.

Dealing with awkward points and comparisons

Deal with counter-arguments and alternative interpretations immediately. You do not want to give them a chance to take hold and distract the reader's mind from your argument. With more complicated counter-arguments, break them down and deal with them point by point. If you just lay out a counter-argument in full followed by your refutation of it, the reader will find it hard to follow and possibly lose track of what you are saying. Similarly, if you are comparing one quite complicated idea or process with another, it is much easier to grasp the comparison if you run the comparisons side by side, point by point.

Deflating your argument

Keep an eye out for constructions that deflate your argument. For example, if you use *definitely* too much you will imply that other points in your argument are not definite. If you write something like 'Personally I think ... ', you imply that other people would disagree with you. Of course if you are stating a personal opinion you should say so.

It is best to start a sentence strongly and then modify it as necessary, rather than starting with a long qualification:

> The result looks promising, although full analysis of the samples has not yet been carried out, and there is some contamination of samples 3, 7 and 15.

Although full analysis of the samples has not yet been carried out, and there is some contamination of samples, 3, 7 and 15, the result looks promising.

The second sentence is much less positive than the first and emphasises the qualifications, rather than that the results look promising.

Distinguishing between fact and opinion

If you are sure of a point in your argument, do not deflate its power by introducing it with phrases like 'In my opinion ...', or 'It is possible that ...' or 'It has been suggested that ...'.

But if you are dealing with points that you are less sure of, or which are just suppositions, let the reader know this – here phrases such as 'It is possible that ...' are appropriate.

18.2 Important Conventions for Writing a Scientific Document

American English and British English

Be aware of the differences between American and British English, most of which are to do with spelling, for example, *grey*, *utilise* and *colour* are standard spellings in British English, but *gray*, *utilize* and *color* are standard spellings in American English. You could use either spelling system providing you are consistent. If you are studying at a British university it is sensible to use British English. Make sure your spell-checker is set appropriately (see also *Appendix 1: Easily Confused Words*).

Capitalisation

The conventions about capitalisation can be contradictory and illogical. Write *I* when writing about yourself, and start sentences with a capital letter. Other places where a capital should be used are detailed below.

Proper names

Always write proper names (*John, Snow Patrol, Gaborone, San Jose, Professor Green,* etc.) with a capital letter. Also capitalise the following:

Nationalities	She is Thai, they are European.
Languages	He speaks Serbo-Croat.
Religions	My family is Buddhist.

Organisations, departments, jobs and areas of work or study

These are normally capitalised when they are used as proper names, the Army, the Labour Party, the Archbishop of York, Your Honour, His Excellency the Jordanian Ambassador, etc.:

Bella Lugosi is Vice-Dean of Para-Psychology at the University of West Cheam. He is in a meeting with the other vice-deans to discuss the recent crisis.

Mr Mercury is Head of Biochemistry at the University of West Cheam. He also teaches music and technical drawing.

so, for example, you would work in a chemistry department, but work in the Chemistry Department of the University of West Cheam.

Names of periods of time and historical events

These should be capitalised if they are used as proper names, for example:

Stone Age, Middle Ages, First World War.

Adjectives and nouns derived from names

These are capitalised if the connection to the name is still felt to be relevant:

Newtonian physics
Roman empire
Aristotelian philosophy

However, we do not capitalise if the connection is felt to be remote:

french windows
arabic script
herculean effort
wellington boot
sandwich
jersey
boycott

or if it is an activity associated with a name:

galvanise
pasteurise

We do not capitalise scientific units derived from names:

kelvin
joule
newton
henry

although we do capitalise the unit abbreviation (see *Appendix 7: SI Units*).

Titles of books and papers

There is a strong convention to capitalise the first letter of the words in titles, except for pronouns, prepositions and conjunctions unless they start the title:

The Adventures of Don Quixote

Sections of text

It is conventional to capitalise, for example, *Introduction* when you are using it to refer to your introduction. Do not capitalise it when referring to introductions in general. Similarly write Table 4, or Figure 5, but talk about figures and tables in general.

Beware of overusing capitals, particularly of CAPITALISING WHOLE WORDS OR SENTENCES. It is hard on the eyes and lacks subtlety. It's a bit like shouting at someone to try and get them to see your point of view: normally, you just irritate them. Look at any printed book and you will see that professional editors use capitals very sparingly.

Remember that abbreviations that are not normally capitalised should never be capitalised in a title or heading; for example, the title 'ANALYSIS OF THE 60 MG SAMPLE' is incorrect. It should read 'ANALYSIS OF THE 60 mg SAMPLE' (because MG means mega-gauss but mg means milligram!). However, titles are better not completely capitalised at all.

Colloquialisms and contractions

Colloquialisms are words or patterns of speech that are used in every-day spoken English. Colloquialisms can easily lead to ambiguity, particularly if your text is being read by a scientist who works in a slightly different field or for whom English is a second language. For example, we recently saw the following in a dissertation:

... the hot DNA probe was hybridised to the filters.

By 'hot' the writer meant radioactive, but there is no reason why the reader should know this. Be especially wary of using colloquial versions of standard abbreviations, such as referring to methanol as MeOH, as biologists do in the laboratory, rather than the standard abbreviation, CH_3OH.

Contractions such as, *it's, there's, I've, lab (it is, there is/are, I have, laboratory)* are colloquialisms which can very easily slip into your text. They should not be there. So run checks for them using the 'Find and Replace' command on your computer.

Definitions

If there is any ambiguity or controversy about any of the words, phrases or ideas you are using, define your terms so the reader knows how you are using them.

Italics

Italics can be used as an alternative to inverted commas to highlight words in a text. Italics are also used to indicate words taken from other languages:

> *a priori*
> *in vitro*
> *in vivo*
> *in silico*

We do not normally italicise 'foreign' words that are in common usage, for example, et cetera, and you would not write 'My friends are on holiday in *Ibiza*'.

Italics can also be used as an alternative to inverted commas for titles in your references:

> Wakeling, D., Ranking, R., Cox, A., and Steele, D. 2016. The mirror in the bathroom. In *Social Psychology in Modern Housing*. Ed. E. Morton (Saxa, Birmingham) pp. 125–137.

Latin names for organisms

Many scientific words are derived from Latin or Greek. They usually do not have an s on the end if they are plural, for example, data, bacteria. Non-English words are often italicised (see *Italics* section, above). Remember that Latin names for organisms are normally italicised. The genus name is capitalised and the species name is not, for example, *Caenorhabditis elegans* or *Arabidopsis thaliana* or *Gallus fritos*.

Laws

Write the names of laws in lower case, for example, the first law of para-psychology, unless the laws contain a person's name, for example Karloff's law.

Literary quotations to preface chapters

Some authors like to put a literary quotation at the start of each chapter. There is no rule against doing this in your thesis, but we do not recommend it as many examiners frown on the practice. We have seen an extreme example in a PhD thesis that not only had literary quotations, but also little illustrations and cartoons in the margins of the first page of each chapter rather like mediaeval illuminations. You would certainly never be failed for this type of style, but not all examiners share your sense of humour. If you do use literary quotations, then make sure that you only use appropriate ones, and always acknowledge the source of the quotation (and check that this reference is correct).

Misuse of words

Be careful of everyday words that have a particular scientific usage, for example, *force, energy, inertia*. Use these words only in their precise scientific meaning. *Parameter* and *variable*, for example, are sometimes confused. *Parameter* means a constant that does not vary, as opposed to *variables,* which do vary. Remember that a datum is something given or assumed as a fact rather than preliminary results or findings – data are things given or assumed as facts. Also remember that *constantly* does not mean *often*, and that *efficient* does not mean *effective*, neither do *varying, variable, varied* and *various* mean the same thing.

Words with similar meanings or spellings are often confused, for example, *to accept* is to receive willingly, *to except* is to exclude or make an exception; *proceed* means to go ahead with, *precede* means to go before. If they are spelt correctly the wrong words will not be picked up by a spell-checker. If you are using a word with alternative spellings, for example *extractable, extractible*, make sure you are consistent. We have listed some problem words in *Appendix 1: Easily Confused Words*.

Typing mistakes can also lead to the wrong word being used. A spell-checker will pick up combinations of letters that do not form any word that it knows of, but your spell-cheekier will knot pique up wards that our correctly spelt butt wrong. It is easy to slightly mistype a word and produce something you did not intend to, particularly words that are anagrams of each other, for example, *sorted* and *stored*. Make sure your text is properly proofread before handing it in.

Numbers

In discursive scientific text, such as your Introduction or Discussion, it is normal to write out numbers from one to ten, for example, *two mice, nine sub-atomic particles*. Fractions of two words can also be written, for example, *three-tenths of a Mars bar*. Larger numbers and fractions, or those written in methods and protocols, and attached to units (such as 28 g or 4 A) are shown as numerals. If you have two numbers together in a sentence, you can write one of them in words to avoid confusion, for example, *seven 5-mm O-rings, fifty-eight 70-year olds*.

It is normal to write numbers in titles, unless it would be very unwieldy:

The Four Seasons
The Seventy Samurai
The Twenty-Five Favourite Sayings of Dr Karloff
The 3,599 Dalmatians
The B-52s' greatest hits
Disco 2000 by Pulp

As far as possible in the discursive sections of your text, such as the Introduction and Discussion, do not start a sentence with a numeral (some disciplines disagree about this, so check a good recent thesis in your field):

24 theories explain the phenomenon of Moog Genesis.

is better written:

Twenty-four theories explain the phenomenon of Moog Genesis

If you are numbering points in your text, put them in brackets. Arabic or Roman numerals are equally acceptable, provided you are consistent:

> A number of things should be taken into consideration when decid-
> ing on the verdict: (1) the sausages were very tasty; (2) the sausages
> were left out on the kitchen table; (3) the defendant is a dog and
> cannot be held to be responsible for his actions.

> A number of things should be taken into consideration when decid-
> ing on the verdict: (i) the sausages were very tasty; (ii) the sausages
> were left out on the kitchen table; (iii) the defendant is a dog and
> cannot be held to be responsible for his actions.

If you are writing out a series of numbers with units, for example, a series of measurements, avoid something like this:

> 12 m, 13 m, 14 m, 17 m ...

it is much simpler to write them thus:

> 12, 13, 14, 17 m ...

If you wish to show a span, such as 12, 13, 14, 15, 16, 17 m, use an en-dash without spaces (see *Punctuation*, below):

> 12–17 m

or write,

> 12 to 17 m.

A colon is used between numbers in proportions or ratios, for example, 5:1.

Referring to people

When writing about people, avoid making generalisations that might cause offence. If you write 'When a nuclear physicist considers this problem he will come to a different conclusion from that of a marine biologist', you are making an unscientific assumption that the nuclear physicist is male. This could cause offence, and, apart from anything else, if your examiner is a woman it might irritate her. You could get round the problem by writing he/she but this looks terrible. Avoid constructions where you are forced to specify a gender unless you know what the gender actually is. A simple way round the problem is to use the term 'they' to mean 'he or she'. 'When nuclear physicists consider this problem *they* will come to different conclusions from marine biologists.'

A bugbear of one of the authors of this book is the use of the word 'man' when 'human' does the same job and has fewer gender-associations. Be warned.

Repeating words

Unless you are consciously doing so to stress or make a point, do not use a word many times in the same paragraph or sentence. It will not affect the meaning of your text, but it is tedious to read and not good writing style. Buy yourself a thesaurus – or check if you already have one in your word processing program. If you find yourself using the same word again and again, look in your thesaurus for an alternative. Before you use an alternative, check what it means. It might only have a similar meaning, which may distort what you are trying to say, or just sound odd:

> While you should use distinctions and definitions where they are needed to avoid confusion, you should snub unnecessary distinctions and definitions.

SI units

SI units (*Système international d'unités*) are the internationally standardised units of measurement. Use them in the correct abbreviated form and only capitalise them if they are normally capitalised (see *Appendix 7: SI Units*). Remember to put a non-breaking space between the number and the unit, for example, 5 kg not 5kg (see *Chapter 13: Numbers, Errors and Statistics*).

Times of the day, days of the week, months and years

Times of the day should be written using the 24-hour clock, 22:13, rather than 10:13 pm. Days of the week are capitalised, for example *Wednesday*, not *wednesday*, and can be abbreviated without a full stop, *Wed*, for example. Months are capitalised, *September* not *september*, and can be abbreviated without a full stop, *Sept*, for example. Seasons, for example, *summer*, are not capitalised. Write years as numerals, 1998, not in full, *nineteen hundred and ninety-eight*.

18.3 The Use of English in a Scientific Thesis or Dissertation

Read published papers and learn to use the conventions of your field so that you can produce a professional-looking thesis or dissertation. Here are some particular points to look out for.

Grammar

Just about everyone makes slips in grammar when they are talking. This is normal and forgivable. However, when it comes to the written word, particularly in a formal document such as a dissertation or thesis, we expect precision and accuracy. Bad grammar is not only irritating to read:

The physicists, what play the piano, is in the next room.

it can also affect the meaning of what you write:

The talented young machine theorist Professor Welch ate an apple roller-blading in Camberwell.

Here it is unclear who was roller-blading, the talented young Professor Welch or the apple. Better construction would clarify the meaning of this sentence.

The talented young machine theorist Professor Welch ate an apple while roller-blading in Camberwell.

In writing, unlike speech, you have time to revise and consider what you are saying, so make good use of this time.

Paragraphs

A lot of people are not quite sure exactly what a paragraph is – or should be. The word paragraph comes from the Greek *para* (a sheet of handwriting) and *graphos* (mark). Originally it was merely a mark in a text to indicate to the reader that they had reached a pause or change of tack in the argument, much as a full stop is there to indicate the end of a sentence. If you play around with your word processing program you can see marks on your screen that indicate the end of a paragraph (for example § or ¶).

There are no rules as to how long a paragraph should be. It depends on the rhythm of your argument and on what looks good on the page. As a guiding principle, try to separate different points into different paragraphs. Newspaper paragraphs tend to be short and punchy, which gives a sense of energy and looks readable in narrow columns. Newspapers are there to communicate facts easily (well, good newspapers are there to communicate facts easily), and you can learn from their style. In your thesis you may not want quite such a frenetic rhythm, but do look at how good newspapers convey information.

Lay your points before your audience with a steady yet varied pace. A short paragraph tends to emphasise the point you are putting across. The pause between paragraphs gives the reader the chance to reflect on, and be amazed by, your insight.

As a general guide, if you are writing in double-spaced type and your paragraph is longer than half a side of paper, it is probably too long. A full page of unbroken text is a daunting sight. The reader is likely to lose the flow of your argument as they wonder when the next break is going to come.

Sentences

A sentence should contain at least one subject and one predicate. The subject is the person, people, thing, or things the sentence is about.

The predicate is something that is said about the subject, and includes a verb. In the following sentence:

I want to go to sleep.

I is the subject and *want to go to sleep* is the predicate. A compound sentence is made of two or more simple sentences, for example:

I want to go to sleep but I have to stay awake.

Complex sentences are made by taking one sentence as a main clause, and adding other sentences in the form of subordinate clauses, for example:

I want to go to sleep, because I am tired, but I have to stay awake.

because I am tired is the subordinate clause.

Sentence length

There are no rules as to how long a sentence should be. Short sentences tend to emphasise a point. However, if you always write in short, basic sentences, your style will come across as abrupt and jerky. You may often have to repeat words and information in order to make your sentences grammatically correct.

On the other hand, the longer a sentence is, the harder it can be to follow the meaning – unless it is well punctuated – and the easier it is to write a bad sentence that just flows on and on for many lines, and has many internal phrases that confuse the reader and usually cause the reader to lose sight of what you were trying to say in the first place, and we have read many examples of these types of sentences in theses and dissertations where we have entirely lost track of what they are trying to tell us because they go on and on for such an enormous number of lines on the page, and in extreme circumstances they can even go on for more than a page, in fact we've both seen sentences in theses that extend and extend for so long that we've entirely forgotten where they started or what they were trying to say to us in the first place. If your sentence is longer than three lines, it is probably too long.

Using nouns as adjectives

Nouns used as adjectives often have a clumsy feel: *apparatus construction*, although using fewer words than *construction of apparatus*, feels ungainly and can be hard to read. This is not to say that nouns should never be used as adjectives. Phrases such as, *hydrogen bond*, and *SI units*, are fine, but try to avoid writing things like *maximum permissible dose administration* or *vegetative propagation measurement*, when you could write *administration of the maximum permissible dose*, and *measurement of vegetative propagation*.

Prepositions

Avoid unnecessary prepositions, such as 'The thesis will be printed out' rather than 'The thesis will be printed'.

Scientific writing often omits prepositions when describing methods or results, so we would write '10 g $CuSO_4$' rather than '10 g of copper sulphate'. We would also write '5 ml water was added' rather than '5 ml of water was added', or '10 nm diameter' rather than '10 nm in diameter'.

Relative clauses

Relative clauses are made using the relative pronouns, *who*, *which*, *that*, etc. For **people** we use the following forms:

Subject	*who*	The woman who arrived yesterday is an engineer
Object	*that, who, whom*	The man who/whom I saw is a biologist
Possessive	*whose*	The person whose car is broken is an astrophysicist

Nowadays 'whom' sounds rather formal. Compare the following sentences:

The child I spoke to is a genius.

The child to whom I spoke is a genius.

You would probably say the first and, although the second is formally more correct, you would probably write the first too.

For **things** we use a different set of words:

Subject	*which* or *that*	The book which/that fell on my foot is heavy.
Object	*which* or *that*	The book which/that I dropped on my foot is heavy.
Possessive	*whose*	The book whose dust jacket I tore is my own.

18.4 Common Problems with Writing

Unnecessary changes of tense or voice in mid-sentence

At times you will have to change tenses in the same sentence to put across the correct meaning, for example:

By the time the apparatus was set up, the chemicals had been prepared and were ready for use.

Here we shift from past to past perfect, which is quite normal. But avoid shifting tenses if it is illogical, or unnecessary:

The purple mouse goes through the maze and learnt that there was a piece of cheese at the end.

We have shifted from the present to the past tense which makes a nonsense of the sentence (see Figure 18.1).

As I continued with the experiment, the formation of gold deposits was observed.

Figure 18.1   The purple mouse went through the maze and learns that the cheese has gone.

Here we have shifted from the active to the passive for no good reason.

Verbs and subjects that do not agree

You have to make sure that the verb you are using is in the right form, or 'agrees with', the subject:

Careful consideration of all the examples are needed.

'are' does not agree with 'consideration'. This should be written:

Careful consideration of all the examples is needed.

In the following sentence *were* does not agree with *the purple mouse:*

The purple mouse, along with the orange mouse, the red mouse, and the blue mouse, were accidentally sent the papers.

This should be written:

The purple mouse, along with the orange mouse, the red mouse, and the blue mouse, was accidentally sent the papers.

even though you would probably say the first version.

Similarly, you should write *there is a number of issues* ... rather than *there are a number of issues* ... even though it can sound a bit strange. If you are unhappy with that, you could always write *there are several issues* ...:

In British English, collective nouns like *committee, group, class,* can be treated as either singular or plural:

The committee was dismissed.

The committee were dismissed.

American English normally treats them as singular.

Each, either, neither, anyone, anything, everyone, everything, no one, nothing, everybody, nobody are all singular so:

Each of the experiments succeeded in their objective.

is wrong – *each* is singular, *their* is plural. The sentence should read:

Each of the experiments succeeded in its objective.

If you are comparing a singular with a plural, the verb will match the one it is closest to:

Either the apparatus or the students are at fault.
Either the students or the apparatus is at fault.

People often have problems with the word *data,* which is not a collective noun, although it is often wrongly treated as one in everyday speech. *Datum* is singular. *Data* is plural. Write, for example, *the data are* ..., *the data show* ..., *the data were analysed,* rather than *the data is* ..., *the data shows* ..., *the data was analysed.*

Similarly, *media* is plural and *medium* is singular, *phenomena* is plural and *phenomenon* is singular, *criteria* is plural and *criterion* is singular. Treating *phenomena* and *criteria* as singular is a particularly common mistake (e.g. *The most important criteria is* ...) and needs to be avoided.

Using the wrong pronoun

> Amy and myself carried out the experiments on 14 September.

> Amy and me carried out the experiments on 14 September.

You would not write *myself carried out the experiments* or *me carried out the experiments* ... The sentence should read:

> Amy and I carried out the experiments on 14 September.

Here is another example:

> The police are accusing Dr Karloff and I of masterminding the Tokyo Affair.

You would not say '... accusing *I*' so you should write:

> The police are accusing Dr Karloff and me of masterminding the Tokyo Affair.

Not completing comparisons

> The blue mouse is larger.

Larger than what? You have to complete the comparison:

> The blue mouse is larger than the elephant.

Also ensure your comparisons make sense, in the sentence:

> The gauze absorbs more water than paper.

It seems as if you are saying that the gauze absorbs some paper but a lot more water. It would be clearer to write:

> The gauze absorbs more water than the paper does.

Dangling participles

These are also known as hanging, unattached and misrelated participles. If they dangle they do not agree with anything, or agree with the wrong subject. This does not normally cause misunderstanding but it can at times, and it is always jarring. For example, if someone was writing about an experiment they carried out using orange mice:

> While conducting the experiment the orange mouse became confused.

This means the orange mouse was conducting the experiment (see Figure 18.2). The participle 'conducting' has to agree with a subject. Here the only subject in the sentence is *the orange mouse*. It would be better to write something like:

> While I was conducting the experiment, the orange mouse became confused.

Figure 18.2 While conducting the experiment the orange mouse became confused.

The split infinitive

The split infinitive is one of the great shibboleths of English. An infinitive is a verb in the form, *to-*, for example, *to go*. To split the infinitive is to put an adverb (a word that describes a verb, such as *boldly*) between the *to* and the verb, for example, *to boldly go* rather than the more orthodox *to go boldly*. Some people say that to split an infinitive is a sin, while others are more relaxed about it (including the *Oxford English Dictionary*). Whatever your views about split infinitives, be careful about using them in your dissertation or thesis. They are more acceptable in spoken than written English. Your dissertation or thesis is a formal document so use them sparingly, if at all. It is possible that your examiners will consider them bad English.

Where to put prepositions

In formal English we put the preposition before the noun or pronoun,

> *At* what are you looking? The radiobiologist *with* whom I am in love.

In informal English the preposition moves to the end of the sentence,

> What are you looking *at*? The radiobiologist I am in love *with*.

As your thesis or dissertation is a formal document it is best to stick to formal English as much as possible. With phrasal verbs, such as 'looking after', you cannot put the preposition before the noun or pronoun, 'The experiment *after* which I was looking', because the sentence then makes no sense, so the preposition has to go afterwards, 'The experiment I was looking *after*'.

18.5 Punctuation

Punctuation can be compared to the stresses, gestures, intonations and pauses that we use in spoken English to make ourselves understood. One usually does not notice good punctuation, but bad punctuation jars and interrupts the flow of your text. It can also affect the meaning of a sentence. Punctuation is used to make the meaning and sense of the writing perfectly clear to the reader and to guide them through it with the minimum of ambiguity. It would be comforting to think that punctuation is governed by a few simple and logical rules. It is not. Punctuation is governed by rules, conventions and taste in roughly equal measures. There is sometimes little logic to it. Generally, the shorter and simpler you make your sentences, the fewer problems you will have with punctuation.

Punctuation marks

Punctuation marks affect the meaning of a sentence: sometimes they help the flow of your text, and sometimes they just litter the page like mouse droppings on a shiny new floor. Use the smallest number of punctuation marks necessary to make your meaning absolutely clear.

Perhaps surprisingly, there are no rules about how many spaces to leave after punctuation marks, but it is conventional to leave a single space after most of them, though it is quite common to put two spaces after a full stop.

The full stop

The full stop (period if you use American English) is used at the end of a sentence. We can also use the full stop when we are laying out lists in the following style:

Common mistakes:
Forgetting to include citations in the references.
Using incorrect journal abbreviations.
Inconsistency of style.
Misspelling authors' names.
Getting the page numbers wrong.

You do not have to use full stops (except after the final point) and could write:

Common mistakes:
Forgetting to include citations in the references
Using incorrect journal abbreviations
Inconsistency of style
Misspelling authors' names
Getting the page numbers wrong.

But, in British English, we should not use them like this:

Dr Karloff's text included many common mistakes: Forgetting to include citations in the references. Using incorrect journal abbreviations. Inconsistency of style. Misspelling authors' names. Getting the page numbers wrong.

Here, semicolons or commas should be used.

If you are referring to lists of items in your text, then always put them into a logical order and keep them in same order throughout your thesis or dissertation, so that you help the examiner to keep track of what you're saying. For example, if you write 'the following sites were tested, Butler, Chassagne, Reed Parry, Gara, Kingsbury', you should always put these in alphabetical order and then use the same order throughout your text i.e. 'Butler, Chassagne, Gara, Kingsbury, Reed Parry'.

The full stop in abbreviations and contractions

In contractions, when we are using simply the first and last letter of the word full-stops are not commonly used: *Mr*, *Mrs* and *Dr* for *Mister*, *Mistress* and *Doctor*. Note that *Ms* is a recent invention and is not an abbreviation for any word and so does not have a full stop after it. For plurals, we also do not add the full stop so *Doctors* would be written as *Drs* without the full stop.

In abbreviations, where we are using just the beginning of the word, it is more common to use the full stop: *Prof.* and *Dept.* for *Professor* and *Department*.

Increasingly full stops are being abandoned altogether in abbreviations and contractions, particularly the more common ones: *TV, UK, MI5, BBC, 1000 BCE, UB40, 11 am, 45 rpm, 70 mph, MSc, PhD, DPhil.* However, use a full stop if there is a possibility of ambiguity. Convention still dictates the use of full stops with some common abbreviations such as *e.g.* (*exempli gratia*, for example), *i.e.* (*id est*, that is), *etc.* (*et cetera*, and so forth), *et al.* (*et alii*, and others), or *viz.* (*videlicet*, namely). Do not use full stops with the abbreviations of SI units.

Whether you choose to use full stops or not in abbreviations, make sure you are consistent. If you are not sure whether or not to use a full stop, err on the side of caution and use one. We came across one brilliant student (first author *Nature* paper and several other papers) who had not put a full stop after *et al.*, and was made, by a particularly pedantic examiner, to correct his thesis by putting a full stop after *every al* in his 300-page PhD thesis (see Figure 18.3).

If you are using a question mark after an abbreviation with a full stop, put the question mark after the full stop, *Do you have a Ph.D.?* This also holds for exclamation marks, *He has not put a full stop at the end of et al.!* However, you should not use exclamation marks in your text because they are too informal! If your sentence ends in an abbreviation with a full stop do not put another full stop in, simply end the sentence with the abbreviating full stop.

If you are using an abbreviation in your text, it is common practice to write the word in full the first time it is used with the abbreviation in brackets after it, and then just to use the abbreviation, but do not do this with standard abbreviations such as SI units, chemical symbols, etc.

Full stops in titles and headings

Do not use a full stop at the end of a heading or title.

Figure 18.3   A quiet night at the Rat and Maze.

The comma

The comma is possibly the most misused of punctuation marks. Sometimes the use of a comma will change the meaning of your sentence, sometimes it will affect the precise shade of meaning, and at other times it is a matter of taste. Used well, commas make it absolutely clear what you mean and give your writing a good rhythm. But, if they are overused, or used incorrectly, they can make your text look fussy, and can, in some cases, cause confusion. We use commas in the following ways.

Commas and conjunctions

Conjunctions are words like *and*, *but*, *therefore*, *so*, used to join clauses together. If you use a comma there is a break between the ideas and the sense of the sentence changes:

> Dr Karloff has gone to Amsterdam or Dushanbe.

> Dr Karloff has gone to Amsterdam, or Dushanbe.

In the first sentence it is equally likely that Dr Karloff went to either place. In the second sentence *or Dushanbe* is an afterthought we do not think as likely as the first possibility. It is more common, but not necessary, to use a comma with conjunctions that indicate a break between ideas:

> The monster was dead and it did not move when Dr Karloff touched it.

> The monster was dead, but it moved when Dr Karloff touched it.

You cannot put a comma between two sentences without using a conjunction, for example:

> Dr Karloff has disappeared, he was last seen catching a taxi to Heathrow airport.

The sentence should read:

> Dr Karloff has disappeared and he was last seen catching a taxi to Heathrow airport.

This is a very common mistake and it is important not to slip into writing this way.

Commas are also used to avoid ambiguities, for example, *however* has more than one meaning. *However* can mean either *nevertheless* or *in whatever way.* You have to be very careful about where you put your commas as it can change the meaning of your sentence:

> Dr Karloff goes to the hotel in Berkeley Square, however Captain Blount waits for him.

Here *however* means *in whatever way* (sitting in the room, hiding in a tree, disguised as a penguin, etc. which does not make sense). In the following sentence:

> Dr Karloff goes to the hotel in Berkeley Square, however, Captain Blount waits for him.

Here *however* means *nevertheless*. You need the comma after *however* whenever you use it in the sense of *nevertheless*. *Still* has similar problems. *Still* can mean *not moving*, *nevertheless*, or it is used to indicate that something is continuing:

> Still I found the exercise useful.

without a comma this either means that you were still whilst finding it useful, or that you continued to find it useful.

> Still, I found the exercise useful.

means 'Nevertheless, I found the exercise useful'. There are other times when a comma is needed to get the right meaning across:

> I turned and examined the samples.

> I turned, and examined the samples.

In the first sentence the samples were turned and examined. In the second sentence you turned and the samples were examined. The commas shows us that *turned* agrees with (goes with) *I*. Keep an eye out for ambiguities like this.

Commas with a list or series of phrases

You can use commas to separate the items in a list or a series of phrases:

> I went to the market and I bought a fat hen, a haddock, an overcoat,
> a photograph of Madness, a lithograph of the Camden canal.

You should use an *and* before the final item in the list:

> I went to the market and I bought a fat hen, a haddock, an overcoat,
> a photograph of Madness and a lithograph of the Camden canal.

People disagree about whether or not to use a comma before the final
and, but it is safer to use one to avoid ambiguities like:

> I went shopping at the shoe shop, book shop, flower shop, super-
> market, delicatessen and hardware store.

Here it is unclear if the delicatessen and the hardware store are two
shops, or one shop that sells fine foods and useful hardware.

You can use commas to separate a series of phrases in the same way:

> In this dissertation I intend to investigate the relationship between
> air pollution and landfish population, the relationship between
> water-borne pollution and landfish size, and the relationship
> between stress and abnormal reproductive behaviour in landfish.

Commas with introductory or inserted words and phrases

Often a comma is not strictly needed, but makes the meaning easier
to follow:

> Finally, we realised we had failed to turn on the centrifuge.

If there is a possibility of an ambiguity, however slight, use a comma
to make your meaning clear:

> Whatever the position of the apparatus I found the experiments
> were inconclusive.

Here it is a possible that you are writing about a piece of apparatus that you found.

Whatever the position of the apparatus, I found the experiments were inconclusive.

Including the comma makes the situation clear.

Commas and relative clauses

We use commas in sentences like:

The blue landfish, *which is highly intelligent,* has escaped and is believed to be responsible for the recent power cut.

or:

The blue landfish, which is highly intelligent, has escaped and is believed to be responsible for the recent power cut, *which has thrown the university into chaos.*

but not here:

The blue landfish *that is highly intelligent* has escaped and is believed to be responsible for the recent power cut.

Commas with if clauses and when clauses

In a sentence like:

You will be ill if Dr Karloff puts the chemical in your tea.

or:

She was ill when Dr Karloff put the chemical in her tea.

we do not need a comma, although it can be useful to use commas to guide the reader through a particularly long and complicated sentence. It is, however, common (though not necessary) to use a comma when the *if* or *when* clause comes first:

If Dr Karloff puts the chemical in your tea, you will be ill.
When Dr Karloff put the chemical in her tea, she was ill.

Inserting words, phrases or clauses into sentences

If you are inserting a clause, phrase or word into the middle of a sentence with commas you must have both an opening and a closing comma just as you would use opening and closing brackets. A very good way of checking to see if you have used the commas correctly is to imagine replacing them with brackets:

Dr Karloff braced himself and, cursing Hornsby, jumped from the open door of the cargo plane.

Dr Karloff braced himself and (cursing Hornsby) jumped from the open door of the cargo plane.

The Dean, Dr Moon, was arrested for drink-driving outside Saltdean Lido.

The Dean (Dr Moon) was arrested for drink-driving outside Saltdean Lido.

E.g., i.e., viz., etc.

We use a comma before and a comma or a colon after words or phrases like: e.g., for example, i.e., that is, and *viz.* to put them in parenthesis:

When served, fried landfish often tastes better with the addition of seasoning, for example, salt or pepper.

Commas and adjectives

Do not put a comma between the last adjective and the noun (i.e. *not* between 'iridescent' and 'gem-stone' in the example below):

> Dr Karloff held up the bright, sparkling, iridescent gem-stone and sighed with satisfaction.

If the adjectives are closely associated forming a kind of compound adjective, the commas are left out, for example, an *air-cooled rotary engine*.

Commas and dates

In British English commas are not put around the year:

> *12 February 1809 was a special day*

in American English they are:

> *February 12, 1809, was a special day*

The colon and the semicolon

The colon

We use the colon between sentences when there is a movement forward in the ideas:

> The chemist and physicist Michael Faraday started work as a laboratory assistant to Sir Humphrey Davey in 1813. In 1821 he began experimenting on electromagnetism, and in 1833 succeeded Davey as Professor of Chemistry at the Royal Institution: he went on to become one of the foremost scientists of his day.

We use the colon to introduce examples and quotations:

> ... many attempts have been made to overcome this problem. For example, Karloff maintains:

The normal phases of catabolism, in which complex substances are decomposed into simple ones, and anabolism, in which complex substances are built up from simple ones, have been bypassed by the technique of Moog Genesis ...

and before a summary of a situation:

The conditions under which I was working changed considerably: my collaborator Adele had left to take up a career on the stage, my supervisor Dr Karloff had disappeared (he is rumoured to be in Cape Town), and I was having severe doubts as to the point of my research.

We can also use a colon instead of a comma after phrases like: e.g., for example, i.e., and *viz.*:

There are two things the Big Bopper likes, *viz.*: Chantilly lace, and a pretty face.

We use the colon before lists of things, for example:

I went to the market and I bought: a fat hen; a haddock; an over-coat; a photograph of Madness; a lithograph of the Camden canal.

Using a colon rather than a comma makes it more list-like. When you are introducing your list with phrases such as *the following, are these, namely*, always use a colon:

I went to market and I bought the following: a fat hen, a haddock, an overcoat, a photograph of Madness, and a lithograph of the Camden canal.

The colon is also used between numbers in proportion, for example, 5:1, and can be used when writing times of the day, for example 21:30.

The semicolon

You can use semicolons to separate clauses which are of more or less equal importance:

> Dr Karloff has disappeared; he was last seen catching a taxi to Heathrow airport.

When using a semicolon in this way you have to make sure that each clause could stand as an independent sentence.

We can also use semicolons instead of commas to separate a series of phrases. They are especially useful when the phrases themselves contain commas:

> The University of West Cheam has recently been involved in a series of unfortunate scandals: the arrest of Dr Karloff on drugs charges; the resignation of Vice-Dean Martin over the Tokyo Affair; the fabrication of student numbers, which only came to light last week, in the Para-Psychology Department; and the failure, despite threats of legal action, to keep up mortgage re-payments for the Salisbury Avenue site.

Apostrophes and inverted commas

The apostrophe

Apostrophes indicate possession, for example, *Keith's ice bucket,* and abbreviations, for example, *He's drunk.*

With names ending in s, it is best to add 's, for example, *Ms Keys's map of Newport,* but with classical names, for example, *Socrates,* it is often omitted, *Socrates' cat.* With plurals ending in s, for example, *thermometers,* you just write *thermometers',* for example, *the thermometers' casings.* Where we have more than one noun we only use the apostrophe with the last noun, *Lorraine and Ricky's wedding.*

Apostrophes can also be used for abbreviations of plurals and numbers, *MSc's, MP's, I went to Tanzania in '15.* Apostrophes are also used in abbreviated words and phrases, for example, *don't, doesn't, haven't*. Do not use these abbreviations in your dissertation or thesis.

A problem, which crops up surprisingly often, is the confusion of *its* and *it's*. *Its* is the possessive. *It's* is an abbreviation of *it is* – do not, of course, abbreviate *it is* to *it's* in your text. And whatever you do, don't write *its'*! This is never correct. It is very easy to slip apostrophes in by mistake and your spell-checker might not catch them, so run special checks for them using your 'Find' or 'Search' command.

Another common misuse of apostrophes is in what is commonly referred to in the UK as 'The greengrocer's apostrophe' for plurals of nouns, for example, *Orange's and Lemon's*. This can often be found in other contexts (*context's?*) now but is ALWAYS WRONG!

Inverted commas

The conventions about inverted commas, or quotation marks, vary from publisher to publisher, we shall follow the conventions of the Oxford University Press. Whatever set of conventions you choose to follow, make sure you are consistent. One point all publishers would agree on is that inverted commas are used to indicate quotes, written or spoken, to show what a person's actual words were. For example, if the actual words spoken were 'I am on the verge of a breakthrough':

The speaker said he 'was on the verge of a breakthrough'.

is wrong. Write:

The speaker said he was 'on the verge of a breakthrough'.

or:

The speaker said, 'I am on the verge of a breakthrough'.

Inverted commas are also used to indicate colloquial or technical expressions, 'yo dude', 'goo-gum oscillator'. Inverted commas can also be used when quoting the title of a publication:

> Pecknold, R., Skjelset, S. (2017). Hunting of landfish by fleet foxes. In 'Country Pursuits', K. Bush (ed) (Orzabal and Smith, Bath) pp. 3–10.

We do not use them for well-known books and publications such as the Bible, Koran, Talmud, Magna Carta, etc.

If you are using a short quote in your main text you can simply put it in inverted commas. It is best to indent longer quotes – but do not use inverted commas with indented quotes.

In British English, if you have a quotation within a quotation, use single inverted commas first, and then double inverted commas for the second quotation. In American English, the convention is reversed.

In British English, if the quotation comes at the end of a sentence, the full stop is put outside the inverted commas; in American English it is put inside. If a punctuation mark is part of the quotation keep it inside the inverted commas, and if the quotation ends in a full stop, question mark, or exclamation mark, you do not need to use another full stop after the inverted comma to end your sentence.

Brackets

In a written text (we are not going to consider their use in equations and formulae) rounded brackets (or parentheses) are used (like a pair of commas) to separate a number, word, phrase or clause from the main sentence. Use them if you are making numbered points in your text:

> We will look at the behaviour of the purple mollusc (1) before, (2) during and (3) after treatment.

You can also use them when you want the effect to be more of an aside than a pair of commas produce:

> The University of West Cheam (formerly West Cheam Institute of Technologickal Arts) has been closed down temporarily.

Do not let a pair of brackets disturb the normal punctuation of the sentence. In the following sentence we need the final full stop:

> The data from the final experiment were inconsistent (see Fig. 6).

We normally only use brackets around short phrases, clauses or sentences we are slipping into a main sentence. Although it is possible to use brackets around whole sentences or groups of sentences it looks very unwieldy. (If you do bracket an entire sentence, the full stop goes inside the brackets, like this.) Square brackets [...] are used mainly with quotations either to fill in some missing information:

> The [principal] aim of the Genetic Engineering Department is to produce blue dandelions.

or with *sic* (so) to indicate a mistake in the original quotation:

> Mr Van Halen wrote, 'We normally put soddum [*sic*] chloride and vinegar, ascetic [*sic*] acid, on our fish and chips'.

Curly (or hooked) brackets {...} can come in handy if you are using brackets within brackets ({...}). However, if your text is getting that complicated (so that you need to have brackets within brackets {like this}) you are probably using too many brackets and should think about rewriting your sentence. Angle brackets ⟨...⟩ are not normally used in written text.

Dashes and hyphens

Normally you don't need to worry about the different lengths of dashes, but sometimes it matters.

There are two sorts of dashes: em-dashes (—) and en-dashes (–). The em-dash (the width of an m) is slightly longer than the en-dash (the width of an n). Hyphens (-) are slightly shorter then en-dashes. You will be able to produce the different marks by playing around with your keyboard (or using the facility to insert special characters). On our keyboards a hyphen (-) is produced by pressing the hyphen key, an en-dash (–) is produced by pressing the CTRL + MINUS SIGN keys together, and an em-dash (—) is produced by pressing the ALT + CTRL + MINUS SIGN keys together (there are also special codes in LATEX which are very simple: - for a hyphen, -- for an en-dash and --- for an em-dash).

Do not put spaces between a hyphen and the words it joins. Opinion is divided as to whether or not to put spaces between an em- or en-dash and the words it separates. The most common practice is to have spaces with an en-dash but no spaces with an em-dash. However, these are not universal rules, so use whichever convention you feel most comfortable with (but make sure you do so consistently).

The en-dash

You use the en-dash (without spaces) to indicate a span of space or time, and page numbers:

> The London–Brighton road.
> 2015–2017
> pp. 1–21

The en-dash is also used to join names that define something, e.g. the Karloff–Welch Law or the Mavers–Murphy Equation.

The en-dash can be used to insert extra information (a sub-clause) into a sentence in the same way as commas and brackets. This is the convention that we have followed in this book, for example:

> The University of West Cheam – formerly West Cheam Institute of Technologickal Arts – has been closed temporarily.

Some word processors will automatically replace a hyphen by an en-dash when it comes between two spaces, which saves having to use the special symbol for an en-dash when typing a sentence like the one above.

Although technically there is a slightly different symbol for the minus sign (−), it is very close in appearance to an en-dash (–) so it is perfectly acceptable to use an en-dash in equations. But don't just use a hyphen in equations, as that is definitely too short, as you can see by comparing these two examples:

$$z = 6x - 4y$$
$$z = 6x\text{-}4y$$

The em-dash

The em-dash can also be used to insert extra information into a sentence in the same way as commas, brackets, and en-dashes (as detailed above) and normally in this case no spaces are used:

The University of West Cheam—formerly West Cheam Institute of Technological Art—has been closed temporarily.

Using an em-dash to insert a sub-clause (rather than en en-dash) is most common in the house styles of American publishing houses and newspapers, but either is acceptable.

You can use an em-dash like a comma before a comment or an afterthought, in which case the effect is more abrupt and immediate:

The data from the final experiment were inconsistent—see Fig. 6.
Multiplying by 5 brings us to the answer—at a price.

You can also use it like a colon:

The conditions under which I was working changed considerably—my collaborator Adele had left to take up a career on the stage, my supervisor Dr Karloff had disappeared (he is rumoured to be in Wellington), and I was having severe doubts as to the point of my research.

In this example we have used brackets rather than dashes to separate *he is rumoured to be in Amsterdam* from the sentence, as further use of dashes would be confusing.

The hyphen

Like the comma, the hyphen is a small and often misused punctuation mark. Hyphens are used to join two words together to make a new word, for example, common-sense. Whether or not words are hyphenated depends on the stage they have reached in their evolution. With fairly recent pairings we tend to write them as two separate words. When they have become reasonably accepted they progress to being hyphenated. Once they have become generally accepted we write them as one word, for example, handkerchief. Although they have been around for a long time, a few words like *common sense/commonsense/common-sense* and *good bye/goodbye/good-bye* still have not settled down into one standard form or another. If you are not sure about a word, be guided by your spellchecker (and dictionary), which will at least be consistent. There are certain times we always use hyphens.

With pure prefixes:

Ex-President, Vice-Dean, Sub-Postmaster

With proper names:

anti-Communist, pre-Christian

With suffixes to single capital initials, symbols, Greek letters, etc.:

X-ray, γ-ray

To avoid double 'i's, 'o's and triple consonants:

anti-intellectual, bell-like, co-operative

With written numbers:

> thirty-five, eighty-seven, three-quarters, five-thousand and twenty-nine

With a series of hyphenated words you can use your hyphens like this:

> We used the three-, six-, and nine-second fuses.

Always use hyphens when there is any danger of ambiguity:

> re-cover, recover
> re-creation, recreation
> un-ionised, unionised

When you are using compound adjectives you must use hyphens.

> Little-used apparatus is a waste of space.

means the apparatus is not used often.

> Little used apparatus is a waste of space.

means the apparatus is little and used. Take care that you hyphenate all the words you need to:

> The anti-animal experiments lobby are picketing the organic sausage factory.

is rather different from:

> The anti-animal-experiments lobby are picketing the organic-sausage factory.

Hyphens can also be used to link two parts of a word split at the end of a line, but it is best to avoid this as it can be confusing and looks bitty. However, note that some word processing packages will do this automatically when it is appropriate.

Apart from after a question mark or exclamation mark – like this ! – you should never use dashes or hyphens next to another punctuation mark. Dashes are not an all-purpose substitute for other punctuation marks. They can liven up your text but do not be tempted to overuse them or your text will look like Morse code.

The slash

This is sometimes called a solidus (/). It used to mean *per*, as in km/h. It should not be used to join two words together. Use a hyphen to do this. The slash is also used to make combinations such as *he/she, his/hers* and *and/or*. We suggest you avoid this if possible as it looks rather clumsy.

The question mark

Keep these for direct questions:

What is the secret of life?

With an indirect questions you do not use them:

In this chapter we will ask what the secret of life is.

The exclamation mark

Over a pint of gin in the bar, you might well say:

Wow! I've discovered a new atom!

There really should not be any place in your dissertation or thesis for exclamation marks (other than as symbols in equations).

Accents

With words borrowed from other languages we keep whatever accents they came with, for example, *précis, pièce de résistance, papier-mâché, façade*.

Keep the accents on people's names when referencing them. It is worth playing around with your keyboard to find out how to get the different accents.

Although you will probably not be using it yourself, you might come across something called a dieresis, which is like the German *umlaut*. It is used to clarify the pronunciation, for example, *naïve* and *coöperation* rather than *naive* and *co-operation* (though *coöperation* is hardly ever used nowadays).

18.6 Common Errors with Punctuation

Fragments

Fragments are constructions like this:

> The blue landfish ate the cheese. Which had been left out by the kindly Mrs Karloff.

Which had been left out by the kindly Mrs Karloff is not a sentence in its own right. The sentence should read:

> The blue landfish ate the cheese that had been left out by the kindly Mrs Karloff.

or,

> The blue landfish ate the cheese, which had been left out by the kindly Mrs Karloff.

Comma splices

Comma splices are constructions like this:

> The yellow landfish ate the blue cheese, it did not like the cheese much.

You cannot simply join two sentences with a comma. Either use a full stop:

> The yellow landfish ate the blue cheese. It did not like the cheese much.

or a semicolon:

> The yellow landfish ate the blue cheese; it did not like the cheese much.

or a conjunction, such as *and*, with or without a comma:

> The yellow landfish ate the blue cheese and it did not like the cheese much.

Run-on sentences

A run-on sentence is something like this:

> Dr Karloff left the country he did not want to be arrested.

Here we have put what should be two sentences into one with no punctuation at all. Again we should use a semicolon:

> Dr Karloff left the country; he did not want to be arrested.

or a conjunction, with or without a comma.

> Dr Karloff left the country because he did not want to be arrested.

Common Mistakes

- Using 'it's' as a possessive instead of 'its'
- Separating sentences with a comma
- Writing incomplete sentences (especially without a verb)
- Using split infinitives

- Using the wrong tense
- Mixing tenses
- Misuse of words
- Vagueness
- Bad punctuation.

Key Points

- Be clear, concise and accurate
- Make sure what you write makes sense
- Make sure what you write means what you want it to
- Proofread you text very carefully, and if possible, ask someone else to look at it too.

Easily Confused Words

With increasing reliance on spell-checkers built into word processing programs, there are many words which will not be picked up if they are used incorrectly. In this appendix we list some of the more common instances of words that can easily be confused either in meaning or spelling.

to accept, to an except

To *accept* is to receive willingly. To *except* is to exclude or make an exception.

acknowledgement, acknowledgment

Acknowledgement is British English; *acknowledgment* is American English.

advice, to advise

You *advise* someone (advise is the verb). They accept your *advice* (advice is the noun).

adviser, advisor

Both spellings are fine, although *adviser* is more common in British English and *advisor* more common in American English.

to affect, to effect, an effect

To *affect* means to act upon something (also, to aim at, and to pretend). To *effect* means to bring about, or accomplish. So, for example:

> Your strange behaviour *affects* your flatmates.

> You *effect* a breakthrough in chemistry.

An *effect* means something caused or produced.

all ready, already

All ready means all of the things are ready. *Already* means by this time.

all right, alright

All right is considered 'more correct', so it is probably best to use this in your thesis or dissertation.

aluminum, aluminium

Aluminum is American English; *aluminium* is British English.

anybody, anyone, anything

These are written as one word, unless you mean any body (just the body), one (thing), or thing.

to appreciate, to understand

Appreciate means to recognise the worth of or increase in value; *to understand* means comprehend.

around

Around means surrounding; do not use it instead of *about*.

around, round

These are interchangeable except in phrasal verbs like 'mess *around*', 'play *around*', and in phrases like 'all year *round*'.

artefact, artifact

Artefact is British English; *artifact* is American English.

as

As can mean since, because or during.

assume, presume

These are more or less interchangeable, although *assume* tends to be used when one is not sure what one is saying is true, and *presume* when one is stating what one believes to be true.

assure, ensure, insure

Assure – to give an assurance to remove doubt.
Ensure – to make certain, to make sure.
Insure – in British English this means to take out insurance; in American English it can also mean the same as ensure.

basal, basic

Basal is used mainly in technical and scientific writing; *basic* in more everyday English.

behavior, behaviour

Behavior is American English; *behaviour* is British English.

below, bellow

Below means underneath; to *bellow* is to shout.

beside, besides

Beside means 'by the side of', or 'in comparison with'. *Besides* means 'in addition to', and 'other than'.

can, may, could, might

Can expresses ability, 'I *can* speak French'. *May* expresses permission and possibility, 'You *may* leave', 'This *may* be the room'. *Could* and *might* can be used either as the past tense of *can* and *may*, or in the present tense to express possibility, 'This *might/could* be the right room'.

cannot

Cannot is usually written as one word. It should not be abbreviated to *can't* in a thesis or dissertation.

color, colour

Color is American English; *colour* is British English.

compare with, compare to

When you want to indicate a similarity between things use *compare to:*

> Shall I *compare* thee *to* a summer's day?

When you want to indicate the differences between things you can use either *to* or *with,*

> *Compared with/to* London, Oxford is a small city.

With is always used when compare is in constructions such as 'Dogs *compare* favourably *with* cats when it comes to loyalty'.

complement, compliment

To have a *complement* means to have a complete set and if one thing (or person) *complements* another, it means that they fit together well. A *compliment* is an expression of praise. A *complimentary* gift is one you did not pay for.

connection, connexion

Both are correct, although *connection* is the more usual spelling in British English.

content, concentration

Content means amount; *concentration* means weight per volume.

continual, continuous

Continual means frequently; *continuous* means without interruption.

could have, could of

'*I could have* come' is good English. '*I could of* come' is nonsense and should never be used.

course, coarse

A *course* is what you study to gain a degree; *coarse* means unrefined.

credible, creditable

Both mean believable, *creditable* also means worthy of praise.

criterion, criteria

Criterion is the singular; *criteria* is the plural.

data, datum

Datum is singular. *Data* is plural. So write *data* in the plural, for example, 'the *data* are ...', 'the *data* show ...', 'the *data* were analysed ...'.

to defuse, to diffuse

To *defuse* is to remove a fuse or to calm down a situation; to *diffuse* is to disperse.

dependent, dependant

One thing is *dependent* on another; *dependant* refers to a person who depends on another person.

different from, different than, different to

All are correct, although some people think that only *different from* should be used. *Different than* is more common in American than British English.

disc, disk

Disc tends to be more usual in British English; *disk* in American English. When referring to computer storage most people write *disk*, as we have done in this guide.

discreet, discrete

Discreet means cautious and circumspect; *discrete* means distinct from.

disinterested, uninterested

Disinterested means impartial, with no interests in the issue. *Uninterested* means indifferent.

to disprove, to disapprove

To disprove means to prove wrong; *disapprove* means to dislike.

effective, effectual, efficacious, efficient

Effective means having the desired effect, coming into operation, or actually rather than theoretically existing.

Effectual means able to achieve the required effect – we only apply it to people in the negative: ineffectual.

Efficacious is used for things (medicines, etc.) and means sure to produce the required effect.

Efficient means doing its job with the minimum of waste.

either/or, neither/nor

Either/or is used for positive comparisons, 'You can have *either* a banana *or* an apple'.

Neither/nor is used for negative comparison, 'He won *neither* fame *nor* fortune'.

elicit, illicit

Elicit means to obtain, for example, *to elicit* information from a supervisor. *Illicit* means unlawful.

eligible, illegible

Eligible means fit to be chosen; *illegible* means unreadable.

to enquire, enquiry, to inquire, inquiry

In British English, *to inquire* and *inquiry* are used for formal investigations, *to enquire* and *enquiry* are used in the general sense. Americans tend to stick to *inquire* and *inquiry*.

farther, further

We use *farther* when talking about distance, *further* when talking about time or degree. 'Aberdeen is *farther* away from London than Cambridge', 'He waited for her a *further* two hours and thought *further* about the terrible secret he had uncovered, then he walked a little *farther*'.

few, fewer, fewest, little, less, least

We use *few* before countable nouns: *Few* people. *Fewer* people. We use *little* before uncountable nouns: *Little* salt. *Less* salt. By *a little* we mean a small amount. By *a few* we mean a small number.

foreword, forward

A *foreword* is an introduction to a book. *Forward* is a direction. The *foreword* is not written by the author of the book, the preface is.

got, gotten

In American English *gotten* is an equivalent of *got*. In British English *gotten* is not used.

gram, gramme

Gram is correct; do not use *gramme*. The abbreviation is simply g with no full stop.

if, whether

If is for possibilities; *whether* is for alternatives, 'I do not know *whether* my grant payment has come'.

impracticable, impractical

Impracticable – cannot be carried out.
Impractical – not practical, not sensible, not reasonable, etc.

into, in to

The student bumped *into* her supervisor, and they walked *in to* discuss the project.

The professors ran *into* the bar, they went *in to* drink a lot of gin and listen to the Stranglers.

its, it's, its'

Its is the possessive of it (the instrument blew *its* fuse); *it's* is only used as an abbreviation of 'it is'. You should not generally use *it's* in a thesis or dissertation. *Its'* is always incorrect.

led, lead

Lead is the element; *led* is the past tense of the verb '*to lead*'.

licence, license

Licence is British English; *license* is American English.

to lie, to lay

Lie, lying, lay, lain – mean to recline, 'The man *lies* down'.

Lay, laying, laid – mean (in non-colloquial English) to place or put something, 'John *laid* the pen on the desk'.

lighted, lit

Both are fine as the past tense of light.

longways, longwise, lengthways, lengthwise

Longways and *longwise* are both possible, but *lengthways* and *lengthwise* are more commonly used.

loose, lose

Loose is the opposite of tight. To *lose* is the opposite of to win (or to mislay).

meantime, mean time, meanwhile, mean while

It is normal to write both as one word, *meantime, meanwhile.*

media, medium

Medium is singular; *media* is plural.

meter, metre

In British English, a *meter* is some kind of measuring instrument, and a *metre* is a measurement of length; in American English *meter* is sometimes used for both. However, *metre* is the internationally agreed term for a measurement of length. Similar considerations to apply to pairs of words such as *center/centre, liter/litre* and *fiber/fibre*.

one, we

One can be used as an indefinite pronoun (also called impersonal and generic pronoun) to mean people in general, '*One* would not like to have to write *one's* thesis in Latin'. We can be used in a similar way, when the writer is including the reader, '*We* need to keep an eye out for spelling mistakes'.

one of the best/worst etc.

This is fine if you mean there are a number of *best* or *worst* things (although one could argue that there can only be one *best* or *worst* thing). It is better to write: one of the better ... one of the worse ...

ought, ought to

Always use *to* with *ought*. 'Dr Karloff *ought to* be here by now.'

owing to, due to

Owing to means because of, '*Owing to* my total lack of experience I set fire to the High Energy Physics Laboratory'. *Due to* means as a result of. It needs a subject + verb in front of it, 'I set fire to the High Energy Physics Laboratory *due to* my total lack of experience'.

per cent, percent, percentage

Per cent is used in British English. American English tends to use *percent*. *Percentage* is a rate or proportion reckoned as so many *per cent*.

phenomena, phenomenon

Phenomena is plural. *Phenomenon* is singular.

to practise, practice

In British English to *practise* is the verb and a *practice* is the noun. American English uses practice for both verb and noun.

to premise, premiss

To *premise* is to assume something from a *premiss*.

premises, premisses

Premises means a building. *Premisses* is the plural of *premiss*.

to prescribe, to proscribe

To *prescribe* is to lay down a rule. To *proscribe* is to ban.

principle, principal

A *principle* is a rule or accepted general truth. A *principal* is either a person in a position of authority or an adjective meaning the most important (referring to a person or a thing). Remember this useful phrase: 'The princip**al** is your **pal**'.

to proceed, to precede

To *proceed* means to go ahead with. To *precede* means to go before.

program, programme

Program is used in American English. In British English we use *programme*, except when writing about computers, when *program* tends to be used (as we have done in this guide).

proof, evidence

Proof is conclusive, *evidence* merely indicates that something might be true. Be very careful about claiming you have proved something.

proved, proven

These are both acceptable as past participles: 'It has been *proved*', 'It has been *proven*'. *Proven* is more often used as an adjective, 'A *proven* idea'.

to quote, quotation

You *quote* a *quotation*.

red, read

'The book was *red*' refers to its colour; 'The book was *read*' refers to what you did with it.

to rise, to raise

Rise, *rose* and *risen* mean to stand up or move upwards – either literally or metaphorically, 'Dr Karloff *rose* to the challenge'. *Raise*, *raised* and *raised*, in British English, usually mean to make something move upwards, 'Dr Karloff *raised* his glass'. *Raise* can also mean to make something grow, 'Dr Karloff has *raised* three landfish from birth'.

sight, site, cite

Sight is the sense. *Site* is a place, as in building site. To *cite* is to quote or reference.

to sit, to set

Sit, *sitting* and *sat* mean to assume or maintain a sitting position, 'I *sat* on the chair'.
Set, *setting* and *set* mean to put something in position, 'I *set* the cheese down on the plinth'.

some time, sometime

Sometime is written as one word except where it refers to an amount of time. 'You should come and spend *some time* in the lab *sometime* soon.'

specialty, speciality

Specialty is American English; *speciality* is British English.

spore, spoor

Spore is a biological unit that can reproduce asexually. *Spoor* is a track or trail left by a wild animal.

stationary, stationery

Stationary means not moving; *stationery* is paper, envelopes, etc.

systematic, systemic

Systematic is used to mean methodical; *systemic* is used as a technical term meaning to do with how a system works.

that, which, who, whom

You can use either *that, which, who* or *whom* in a *defining* relative clause (without commas):

'The landfish *which/that* were put in the refrigerator have turned blue.'

'The woman *who* ate the mouse is in the cupboard.'

'The man *who/whom* Dr Karloff shot has recovered.'

You can use *who, whom* or *which* with *non-defining* relative clauses (with commas):

'The landfish, *which* were put in the refrigerator, have turned blue.'

'The woman, *who* ate the mouse, is in the cupboard.'

'The man, *who/whom* Dr Karloff shot, has recovered.'

times, fold, per cent

'*Times*', '*fold*' and '*per cent*' are often confused. We use them in the following ways:

'In the second experiment we used three *times* the concentration of NaCl' but, 'In the second experiment the concentration was increased *threefold*'.

If you write 'In the second experiment the concentration was increased three *times*' you mean you increased the concentration on three different occasions, but give no information as to by how much you increased it. You could also write 'In the second experiment the concentration was increased by 200 %'.

to, too, two

To is the preposition; *too* means 'also' or indicates 'excess' as in '*too* much'; *two* is the number.

toward, towards

Both are acceptable.

try to, try and

The words 'to try and …' are very commonly misused. 'To *try and* visit' means to try (to do something) *and then to visit*.

'To *try to* …' means just what it says. 'To try to visit', means … er … to try to visit.

We have the same problem with *go and*, *come and* etc., which should be written 'go to …', 'come to …', etc.

up to date, up-to-date

Compound phrases like this need only be hyphenated when used as an adjective. For example, it is important to keep *up to date* with research in your field as this ensures you have *up-to-date* information in your thesis. In the later example 'up-to-date' is a compound adjective that describes 'information'.

used to, use to

'*I used to* play rugby at school.' '*I use* a pen *to* write my name.'

what ever, whatever, who ever, whoever, how ever, however, etc.

Write these as one word except in questions where the *ever* is used for emphasis: '*What ever* do you think you are doing?'

whose, who's

Whose is the possessive of who; *who's* is an abbreviation of who is, which you should not be using in your thesis or dissertation.

worth while, worth-while, worthwhile

Normally this should be written as *worthwhile*.

Prefixes and Suffixes

Sometimes prefixes will be joined to other words by hyphens, but usually this is only necessary to avoid ambiguity, for example, to distinguish between un-ionised and unionised – see the section on hyphens in *Chapter 18: The Use of English in Scientific Writing*.

Prefixes

anti-, ante-

The prefix *anti-* means against; *ante-* means before.

extra-

Extra- means outside, for example, *extra-curricular* activities are those undertaken outside that which is taught as part of a curriculum or course.

inter-, intra-

The prefix *inter-* means between, among; *intra-* means within or on the inside. Hence the distinction between an institution's intranet and the global internet.

Suffixes

-able, -ible

Your spell-checker will pick out any misspellings you make of words that should take either -*able* or -*ible*, but there are some words that take different meanings in -*able* and -*ible*, for example:

> contractable – liable to be contracted as a disease or habit
> contractible – capable of contracting or drawing together
> forceable – able to be forced open
> forcible – achieved by force

other words can take either -*able* or -*ible*, for example:

> collapsable, collapsible
> extendable, extendible
> extractable, extractible

Another quirk with -*able* is whether or not to omit the 'e' in some words such as *nameable* or *namable*, *tunable* or *tuneable*, *likeable* or *likable*, or *useable* or *usable*. Usually either is strictly possible, but often one spelling is more common than the other. Be guided by your dictionary and spell-checker.

-ae, -as

Most words ending in 'a' have a Latin root and many of them strictly take '-*ae*' as the plural, for example, *larvae*. However, as our language evolves we tend to make a plural by putting an s onto the end of many of these words, for example, the plural of *nebula* is written as *nebulae* or *nebulas*. Some words now always end in an 's' if plural, for example, *areas*. Words such as *media* and *data* are already plural. If in doubt, consult a dictionary.

-ative, -ive

Most words take the *-ative* suffix, for example, *representative* rather than *representive*. Be guided by your spell-checker and dictionary.

-ic, -ical

When making an adjective from a noun, some words always take *-ic*, e.g. *alcoholic*, some words always take *-ical*, e.g. *chemical*. Some words use either *-ic*, or *-ical*, but have different meanings, e.g. *economic*, *economical*. Some words differ between British and American English, for example, British English prefers *-ical* (*geological*), American English tends to prefer *-ic* (*geologic*).

-ion, -ment

Some words take *-ion* as a suffix, some take *-ment*. A few words take both but have different meanings, for example, *excitation* and *excitement*. Be guided by your dictionary.

-ion, -ness

Some words take *-ion* as a suffix, some take *-ness*. Some take both but have different meanings, for example *abstraction*, *abstractness*, *correction*, *correctness*. Be guided by your dictionary.

-ist, -alist

Some words can take either, for example, *educationist*, *educationalist*, *horticulturist*, *horticulturalist*. Be guided by your dictionary and spell-checker.

-ise, -ize

Some words always use *-ise*, such as *advertise*, *advise*, *hybridise*, whereas other words take either *-ise* or *-ize* in British English.

However, just to add confusion, some people and organisations in Britain adopt the 'Oxford English' convention, which always uses-*ize* rather than -*ise*. American English always goes for the -*ize* option, for example, *hybridize*. Be consistent with which one you use; if you are submitting a thesis or dissertation to a British university or college, it is probably wisest to use -*ise* (as we have done in this guide).

-or, -er

Some words end in -*or*, some words end -*er*, for example, *teacher* and *distributor*. Some can take either, for example *advisor*, *adviser*. On the whole, Latin-based words tend to end in –*or*; be guided by your spell-checker and dictionary.

-os, -oes

There are no universal rules as to whether all words ending in *o* should have an -*oes* ending in the plural or an -*os* plural, so we get *potatoes*, and *Eskimos;* be guided by your spell-checker and dictionary.

Wordy Words and Phrases

It is very easy for wordy words and phrases to slip into your text, but they are usually superfluous, so try to remove them if you can. Ask yourself if they add any extra meaning to what you have written. If not, take them out!

Our current most-wordy phrase comes from a thesis one of us recently marked. The student wrote,

> ... this could actually be a consequential resultant rather than the causative element.

Crikey. What they meant was, '... this could be a result, rather than a cause'.

If you were an examiner, which phrase would you rather read? And which phrase would leave you reaching for a stiff gin and tonic?

above

Above can cause wordiness, for example,

> The experiment outlined *above* proved crucial to our research.

You could just write,

> This proved crucial to our research.

Below is a little more useful. You can use it as a signpost for related information or argument the reader will find useful, for example, *I will deal with this contradiction below.*

all, all of

Try to get rid of the '*of*': '*all the samples*' is more concise than '*all of the samples*'.

amount

'A large' *amount* is vague, and 'a maximum' *amount* is redundant. Maximum and minimum are amounts.

area

> My research was in the *area* of delta wing aeronautical design.

This could be written more concisely,

> My research was in delta wing aeronautical design.

as far as

> *As far as* bore holes are concerned, I had no difficulties.

You could just write,

> I had no difficulties with bore holes.

both, both of

> *Both of* the traces indicated an increase in activity.

This does not need the *of* and could be written,

> *Both* traces indicated an increase in activity.

capability

> The plastic has the *capability* of reforming itself.

You could just write,

> The plastic can reform itself.

cause and result

At times you have to make clear what is a *cause* and what is a *result*, but there are often quicker ways of saying what you mean; compare the following, for example,

> The addition of NaCl *caused* an improvement in taste.
> The addition of NaCl improved the taste.

clearly demonstrates, shows

If data *clearly demonstrate* a phenomenon, then they really *show* it.

> The spectrophotometer readings *clearly demonstrate* a decrease in density.
> The spectrophotometer readings *show* a decrease in density.

definitely

Unless there is some question hanging over what you are saying, *definitely* is unnecessary and often looks desperate to be convincing.

due to the fact that

This really means 'because' ...

furthermore

This is a rather vague word that comes at the start of a sentence and means something like 'in addition'. If you find yourself using it, you should ask yourself whether it is actually adding anything to your text.

in colour, in appearance

Both of these phrases are unnecessary,

> The landfish was red in colour.

> The landfish was red in appearance.

This could simply be written as,

> The landfish was red.

in order to

Use 'to'.

literally

A word best avoided in your thesis. You whole text should be literal.

manner

Manner always makes your sentence wordy.

> The chemicals were added in a slow *manner*.

> The chemicals were added slowly.

moreover

This is a slightly fluffy word that means something like 'besides'. If you find yourself using it, you could well have it in your text for no reason other than to give you a mental pause. It is likely to irk your examiners unless you really need it there. (It certainly irks one of the authors of this book.)

nature

This is often superfluous, for example:

> Chewing gum has an elastic *nature*.

> Chewing gum is elastic.

personally

> *Personally* I think ...

This is not only wordy but can deflate your argument. It implies that other people would disagree with you. *I think* does the job better. Alternatively, if you are writing an opinion in your own thesis or dissertation, the examiners know this is what you think, so just write the opinion.

process

Unless you are actually discussing a *process* this is a word to avoid. For example, you do not have to write 'the stratification *process*', you can just write 'stratification'.

pooled together

If you pool samples, then they must be together, and the 'together' is redundant, so you could write,

> The sulphide samples were pooled together.

as simply,

> The sulphide samples were pooled.

reason, because

You only need one of them, as they do the same job.

> The reason the temperature rose is because the heater was turned on.
> The temperature rose because the heater was turned on.

seldom ever

You do not need the *ever*.

similar, very similar

The *very* is redundant. Writing that two items are *very similar* tells us no more than if we write they are *similar*.

sized

> Large *sized* ...

If something is large, we know this refers to size, so the word *sized* is redundant.

that

One *that* in a sentence is normally more than enough unless you want to stress a point; for example, the first and second sentences are clear, the third sentence is full of redundant *thats*:

> I found I could not move, had dropped my samples, felt ill, had a headache, and had fallen down the stairs.
>
> I found *that* I could not move, had dropped my samples, felt ill, had a headache, and had fallen down the stairs.
>
> I found *that* I could not move, *that* I felt ill, *that* I had dropped my samples, *that* I had a headache, and *that* I had fallen down the stairs.

the field of

This is useful when you are giving the general area in which you work,

My research was in *the field of* para-psychology.

but it can get overused,

The field of para-psychology is seldom taken seriously by other scientists.

This could equally well be written,

Para-psychology is seldom taken seriously by other scientists.

Words that Cause Vagueness

Here we provide a list of some common words to avoid in scientific writing, and comparison words for which the values being compared must be defined. This list is not exhaustive, but it contains common words you should look out for in your text and avoid using because they are often inappropriate in a scientific document, although they may be great! or marvellous! in every day conversation.

If you use words that give comparisons, such as higher than/ lower than/wider than/greater than ... etc., you need to state what the value is being *compared to* in a way that tells the reader real amounts – 10 miles or 10 mm, for example.

Quality and Manner

brilliant
excellent
fantastic
good
great
lovely
marvellous
wonderful

awful
bad
dreadful
terrible
worse

beautifully
better
brilliantly
excellently
fantastically
marvellously
well
wonderfully

awfully
badly
dreadfully
terribly

Temperature

cold
cool
hot
warm

Size

big
high
huge
large
little
long
narrow
short

small
thin
tiny
wide

Amount

a few
few
least
less
little
many
most
more
much
some

Speed

fast
quick
quickly
speedily
swift
swiftly
slow
slowly

Time

at once
lately
recently
soon

Frequency

frequently
hardly ever
occasionally
often
periodically
rarely
scarcely
seldom
sometimes
usually

Degree

almost
barely
considerably
enough
extremely
fairly
far
greatly
hardly
highly
little
nearly
negligibly
poorly
quite [unless you mean completely]
rather
really
scarcely
so
too
very

Latin Words and Abbreviations

We have given the meanings of some words, phrases and abbreviations that you may come across in scientific as well as everyday writing. We have not always given the literal translation; sometimes we have given the current day meaning. Plain English is best for most writing, including scientific writing, so do not use these words and phrases unless they are appropriate. Non-English words are usually italicised, apart from those in common use, for example, generally *in vivo* would be italicised, but etc. would not.

ab extra	from the outside
ab initio	from the beginning
	'The project has been under-funded *ab initio*, and therefore has gone slowly.'
ab intra	from within, from the inside
addendum	a thing to be added
addenda	the plural of addendum
ad hoc	to this, for this specific purpose
	This refers to something temporary, something just used for a specific and time-limited purpose. 'An *ad hoc* election was held and Jack and Megan were voted in as the graduate students in charge of fetching the beer at the meeting.'

ad infinitum	forever
	'Margaret felt as if her chemistry tutorial stretched on *ad infinitum*; she resolved to study law instead.'
am, a.m.	morning
	Ante meridiem, this actually means before noon.
a priori	from what is known
	'We have no *a priori* knowledge of how many genes are involved in Down syndrome.'
appendix	something added
appendices	the plural of appendix
bona fide	in good faith, honestly, the real thing
	'He is a *bona fide* representative from the chemicals company, and not a Moog Genesis salesman.'
c., ca., *circa*	about
	Use *circa* when you are not sure about a date. 'We believe that the Head of Department was born c. 1930.'
cf.	compare
	cf. is an abbreviation for *confer*, which is Latin for compare. 'I think Dr Fleetwood's drum solos are the best in the Department (but *cf.* Dr Copeland's drum solo at the Christmas party).'
corrigendum	an item to be corrected
corrigenda	items for correction
	You may come across a *corrigenda* when a paper is being corrected at the proof stage. It is simply of a list of things to correct.
de facto	in reality
	'Dr Shocked is the *de facto* Head of Department, because her boss, Professor Albarn, spends so much time bird-watching.'

de novo	from new, afresh

'We were not happy with the quality of the phosphoglycerol supply and so we decided to synthesise it again, *de novo*, from new stocks of chemicals.'

diem perdidi	another day wasted

Actually, you are unlikely to ever see this written down in anything scientific, but we all know the feeling.

e.g.	for example

This is an abbreviation of *exempli gratia*, which literally means 'for the sake of example'. The abbreviation, e.g., is not normally italicised.

ergo	therefore

These days it is not normal to use this in a scientific text.

errata	list of errors

You may encounter such a list when correcting a scientific paper. *Errata* is simply the plural of *erratum*, meaning error or mistake.

et al.	and others

Et al. is an abbreviation for either *et alii*, or *et aliae*, or *et alia*, which mean and other men, women, and things, respectively. Use *et al.* to mean 'and others' and so avoid long lists, such as lists of authors. 'The paper by Geldorf *et al.* is the result of a collaboration between 30 different laboratories.'

etc.	the rest, and so on

This is an abbreviation for two words, *et cetera*, which mean 'and the next'.

ibid., ibidem   in the same place

You are unlikely to come across this in scientific writing, but we have included it just in case. It means refer to the identical source as the previous one, in other words, look at the same reference as has just been cited.

id., idem   the same

Again, this is unlikely to crop up in scientific literature, but just so that you know, when an author is cited many times, the abbreviation *id.* is used in place of the author's name, after the first citation of that name.

i.e., *id est*   that is

'Their white lab coats were stained with loading dye, i.e., the stains were blue.'

infra   below, underneath

Compare infrared with ultraviolet.

in silico   by computer,

The genes were found *in silico* by analysing the DNA database with the latest software.

in situ   in the natural location

In situ is generally used to indicate that the material you are working with was studied in its natural location. '*In situ* hybridisation techniques allow us to visualise the chromosomes within a cell nucleus.'

inter   among, or between

Use inter to refer to something between individual entities. 'There is a lot of inter-university rivalry between the Oxford and Cambridge boat race teams.' 'The intermolecular distances are very high because the molecules repel each other.' Do not confuse inter and intra. Inter is normally used as a prefix, and is not italicised.

in toto	totally, completely
	'The cost of the analyses, including all extras, comes to £28,000 *in toto*'.
intra	within
	Use intra to refer to something within an individual entity. 'Intramolecular forces keep the atoms together within the crystal.' 'It is sometimes hard to find polymorphisms from two strains in an intra-species cross.' 'Christmas can be a chore because of intra-family rows.' Do not confuse inter and intra. Intra is normally used as a prefix, and is not italicised.
in utero	in the uterus
	This phrase is refers to the embryo or fetus in the uterus. '*In utero* surgery on the foetus is hard to perform but can correct some disorders.'
in vacuo	in a vacuum
	This phrase is most often used literally by physicists, chemists and engineers, and can also be used metaphorically by the rest of us to mean 'in isolation' – avoid this usage in your text.
in vitro	in the test tube (or your equivalent)
	This phrase is literally translated as 'in glass'. We use it for when referring to studies taking place in glass (or plastic) containers.
in vivo	in the living organism
	'Chrissie and Jim are studying the depolarisation of the heart, *in vivo*.'
locus	place
	'The gene maps to this locus on the chromosome.'

N.B.	take note
	N.B. is an abbreviation of *nota bene,* which means 'note well'. This abbreviation should not find its way into your writing because it is too informal, and suggests that you have not properly explained something.
non sequitur	it does not follow
	A non sequitur is a conclusion that does not logically follow from the given premiss.
p.a.	per annum, each year, yearly
	'The bench fees are £15,000 p.a.'
per capita	per head, individually
	'The charge, per capita, for the lab's day trip to Brighton is £65.'
per diem	daily, each day
	'The charge, *per diem,* for Professor Finn's stay in Te Awamutu is £180.'
per se	by, or in itself
	'I am not discussing professors *per se,* I am talking about academics in general.'
pm, p.m.	afternoon
	Post meridiem, actually means after noon.
post mortem	after death, autopsy
	'Dr Cooper performed a post mortem on the body.' Post mortem is a phrase in common usage and does not need to be italicised.
post partum	after parting, immediately after childbirth
	'Isolation in a high rise flat may exacerbate post partum depression.'

QED

thus it is demonstrated

An abbreviation for the Latin *quod erat demonstrandum*. Do not use this in your scientific writing, it is archaic and very few people will know what you mean.

q.v.

which see

Quod vide, q.v., is a way of referring you elsewhere in the text. 'A cow is a large quadruped *(q.v.)*.' This sentence is telling you that you can look up the word 'quadruped' in the text if you need to.

sic

so, thus

Sic is used to draw attention to a mistake in the original source that you are quoting. This can either be a spelling mistake, a factual mistake or a mistake in usage. *Sic* is usually placed inside square brackets. 'The newspaper article stated that domestic salt was soddum [*sic*] chloride and vinegar was ascetic [*sic*] acid.' 'He wrote that crossing the dessert [*sic*] by camel was arduous.' 'The travel guide said the Scottish [*sic*] group the Stereophonics sounded different from the Welsh [*sic*] group called Kasabian.'

status quo

the state things are in

Status quo refers to the current conditions. 'By paying scientists higher salaries we may alter the *status quo* of society.'

ultra

beyond

Compare ultraviolet with infrared.

versus, vs., v.

against

'Today's football match is the Physics Department vs. the Biochemistry Department.'

vice versa   the other way round, the position being reversed, conversely

'Dr Hartman was an excellent collaborator for Dr Johnston and Professor Simmons, and vice versa.'

viva voce   oral examination

Viva voce literally translates as 'with the living voice', but has come to refer to an oral exam.

viz.   namely

Viz. is an abbreviation of *videlicet*, which literally translates as 'it is permitted to see', and is used to mean 'namely'.

The Greek Alphabet

capital	small letters	name
A	α	alpha
B	β	beta
Γ	γ	gamma
Δ	δ	delta
E	ε	epsilon
Z	ζ	zeta
H	η	eta
Θ	θ	theta
I	ι	iota
K	κ	kappa
Λ	λ	lambda
M	μ	mu
N	ν	nu
Ξ	ξ	xi
O	o	omicron
Π	π	pi
P	ρ	rho

capital	small letters	name
Σ	σ, ς	sigma
T	τ	tau
Y	υ	upsilon
Φ	φ	phi
X	χ	chi
Ψ	ψ	psi
Ω	ω	omega

Appendix 7

SI Units

The SI units (*Système international d'unités*) are the internationally standardised units of measurement. The system has seven base units (m, kg, s, A, K, cd, mol) and two supplementary units (rad, sr). All the other units are derived from these.

Each unit has a symbol. The symbols are not capitalised unless the measurement is named after a person, in which case the first letter of the symbol is capitalised. The names of the units are never capitalised. Do not use full stops with the abbreviations of SI units. Multiples of the units are given in decimals (for example, micro 10^{-6}, milli 10^{-3}, kilo 10^3, mega 10^6). The only commonly used decimals are those that are a multiple of three. Remember to use the correct abbreviations for the multiple, for example 'kilo' is abbreviated by a lower case k, not and uppercase K, similarly mega is M and milli is m.

Note that kelvin and K are not used with the word 'degree' or the degree symbol (°).

SI Base Units

Name		Symbol	Abbreviation
length		metre	m
mass		kilogram	kg
time		second	s
electric current		ampere	A

thermodynamic temperature	kelvin	K
luminous intensity	candela	cd
amount of substance	mole	mol
plane angle	radian	rad
solid angle	steradian	sr

Derived Units

absorbed dose	gray	Gy
activity	becquerel	Bq
dose equivalent	sievert	Sv
electric capacitance	farad	F
electric charge	coulomb	C
electric conductance	siemens	S
electric potential difference	volt	V
electric resistance	ohm	Ω
energy	joule	J
force	newton	N
frequency	hertz	Hz
illuminance	lux	lx
inductance	henry	H
luminous flux	lumen	lm
magnetic flux	weber	Wb
magnetic flux density	tesla	T
power	watt	W
pressure	pascal	Pa
volume	litre	l,L

Multiples to be Used with SI Units

multiple	prefix	symbol
10^{-24}	yocto-	y
10^{-21}	zepto-	z
10^{-18}	atto-	a
10^{-15}	femto-	f
10^{-12}	pico-	p
10^{-9}	nano-	n
10^{-6}	micro-	μ
10^{-3}	milli-	m
10^{-2}	centi-	c
10^{-1}	deci-	d
10^{1}	deca-	da
10^{2}	hecto-	h
10^{3}	kilo-	k
10^{6}	mega-	M
10^{9}	giga-	G
10^{12}	tera-	T
10^{15}	peta-	P
10^{18}	exa-	E
10^{21}	zetta-	Z
10^{24}	yotta-	Y

Non-SI Units

Of course many other units are used for historical or practical reasons in different fields. Some examples are given below. In general it's fine to use these units if they are the conventional ones to use in your field, but if there is a suitable SI alternative, consider whether it would be better to use that. Also, if there is any doubt about whether your readers will know what the unit is, give a definition the first time you use it.

Quantity	Unit	Definition
atomic and molecular masses	Dalton (Da)	1 g/mol
base pairs (DNA)	bp, kb, Mb	1, 10^3, 10^6 base pairs
computer memory	Megabyte (MB)	Approx. 10^6 bytes (also GB, TB etc.)
length (astronomy)	astronomical unit	1 AU is equal to the Earth-Sun distance (approximately 1.49×10^{11} m)
length (astronomy)	parsec	1 pc = 3.09×10^{16} m
magnetic field strength	gauss (G)	10^{-4} tesla
power (electronics, acoustics)	decibel (dB)	Logarithmic unit: 10 dB is equal to a factor of ten in power
pressure	Torr	Approx. 130 Pa (much better to use mbar or Pa)
pressure	mbar	Approx. 100 Pa
temperature	degrees Centigrade (°C)	The Centigrade temperature value is the value in kelvin minus 273.15
wavelength	Ångström (Å)	0.1 nm

Alternative Systems of Units

SI units were introduced to give consistency in the use of physical units across all fields. The SI was based on the previous MKS system (which used the metre, kilogram and second as its three base units).

The earlier CGS system (based on the centimetre, gram and second) had a different approach to electrical and magnetic units.

In some fields scientists still use conventions where the standard SI units are replaced by a set of alternative units that have more appropriate sizes for that particular field, or they even drop units altogether by setting a number of fundamental constants equal to unity. Two particular examples of this are given below.

Atomic units

Here the units are chosen to be appropriate for atomic physics. This is done by defining the fundamental constants to have values such that units relate to the properties of the hydrogen atom. The electron mass (m_e), the elementary charge (e), the reduced Planck constant $(\hbar = h/2\pi)$ and the Coulomb constant $(1/4\pi\varepsilon_0)$ are all set to unity. Then the atomic unit of length turns out to be the Bohr radius, the speed of light is $1/\alpha = 137$, and the atomic unit of energy is twice the binding energy of hydrogen in its ground state. The abbreviation used for all these units is 'au' or 'a.u.'. When working in atomic units, all equations become simpler: they no longer contain constants such as e and m_e because these have been set to be equal to 1.

Natural units

Natural units are used mainly in the areas of cosmology and particle physics and again the benefit is that many equations become much simpler by defining some of the fundamental constants to be equal to unity. The definition of natural units is that the speed of light (c), the reduced Planck constant $(\hbar = h/2\pi)$ and the Boltzmann constant (k_B) are all set equal to unity. Additionally the energy scale is usually defined by defining the energy unit to be the electron-volt (eV). (Alternatively, for Planck units the gravitational constant (G) is set equal to unity.)

Arbitrary units

Sometimes graphs are labelled in 'arbitrary units' where it is not possible to be exact about the unit – for example, if a reading is taken from an uncalibrated scale – and it is not important to know the absolute value. To avoid confusion, the axis should be labelled 'Arbitrary Units' or 'Arb. Units' but not 'A.U.' or 'AU', which can have other meanings.

Index

Printed in the United States
By Bookmasters